Foreword by Susan Stamberg

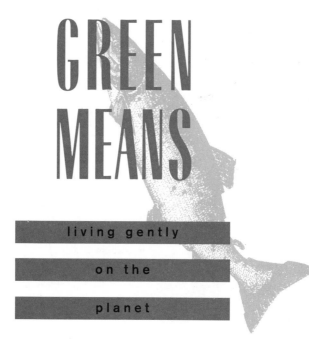

GREEN MEANS

living gently

on the

planet

Aubrey Wallace

As Seen on Public Television

KQED
BOOKS

SAN FRANCISCO

Publisher: *Pamela Byers*
Project Editor: *David Gancher*
Cover and Book Design: *Amaryllis Design*
Printing services: *Penn&Ink*

For KQED:
President & CEO: *Mary G. F. Bitterman*
Vice President for Publishing & New Ventures: *Mark K. Powelson*

Copies of this book are available to educational institutions,
environmental groups, and public broadcasters at attractive quantity
discounts. Contact KQED Books & Tapes, 2601 Mariposa St.,
San Francisco, CA 94110.

Library of Congress Cataloging-in-Publication Data
Wallace, Aubrey.
      Green Means: living gently on the planet/Aubrey Wallace.
      p.   cm.
      Includes bibliographical references.
      ISBN 0-912333-30-8 (pbk.): $9.95
      1. Environmental education—United States.
      2. Environmentalists.
      3. Green Means (Television program)   I. Title.
GE80.W35 1994
94-37998
363.7—dc20
      CIP

ISBN 0-912333-30-8

Manufactured in the United States of America
10 9 8 7 6 5 4 3 2 1

Distributed to the trade by Publishers Group West

This book was printed on acid-free paper with soybean-based inks.

CONTENTS

Foreword by Susan Stamberg

Introduction by Peter Stein

FOREWORD

Susan Stamberg

The taxi was immaculate and perfectly air-conditioned. Tayo, the polite and well-spoken driver, knew where he was going (rare, in Washington, D.C.), and was extremely good-natured (rarer still). The ride was even more unusual because of what was fueling it. A small, green-glowing glass button on the dashboard gave a clue. It signaled that the car was running on nonpolluting, energy-saving natural gas. Tayo didn't mind Vince the cameraman and all his gear, crowding next to him in the front seat, backwards, his lens pointed at my face. Tayo was having fun. So was the cameraman. So was I. The second season of *Green Means* was underway.

The pollution-free, environmentally conscious journey this book represents began in 1992, when KQED executive producer Peter Stein and his band of green-minded television pals began working on a series of brief (five minutes, mostly) profiles of environmental heroes—men and women doing their very best on behalf of the planet, trying to make it a green and nourishing place for all living things. The projects and efforts are heroic, but these are very ordinary people. They're farmers, fisher folk, students, teachers, and foresters. They've affected rural and urban areas on several continents. Often, they have recruited volunteers to help in the greening process. Their stories make us think about their lives—and wonder how we might mirror their commitments in what we do. As host of the television series, I helped to tell their stories on the air.

Sally Fox stays with me just about every time I go shopping for something new to wear. Sally grows colored cotton. That's right, colored cotton. Earth tones, of course—beiges, browns, and green. That way, she avoids the harmful bleaching and dying process that turns traditional white cotton into colored cloth. It's a multimillion-dollar business now (although Sally certainly had her doubts when she started out). On *Green Means*, and in these pages, we report how she did it.

As a city person, I confess that I was tempted to think that environmental concerns were best addressed by rural people, until *Green Means* presented some fertile ideas from New York City and Newark. Staten Island—one of New York's five boroughs—is the site of a salt-marsh restoration project. Snowy egrets, glossy ibis, black-crowned night herons (such wonderful names!) nest near some of the worst urban blight in the country. The birds are being protected by the community; neighborhood school children pitch in to help clean up and plant. And Newark has the most comprehensive recycling program on the East Coast. Again, young people are involved—helping to separate the glass from the paper. Some one hundred local businesses are also involved in the recycling of products. The city has put its muscle behind the project: you get a ticket and have to pay a fine if you mix cans and bottles with garbage. In Newark, they're making sure the Garden State honors its nickname.

These are just a few of the stories we tell on *Green Means*. We also offer specific tips—things all of us can do to green up our world. Some of the tips can be applied on shopping trips—looking for meat and eggs that have been produced in an environmentally conscious way, selecting the right wool (and Sally's cotton), finding organic ways to control garden pests, saving precious resources like water by letting your lawn grow shaggy (and then putting the cut grass into compost piles). It's easy to shop more greenly—reading the labels and ruling out harmful ingredients. Even something as simple as getting a leaky faucet fixed can save gallons of fresh water. These pages are full of plain advice like that—obvious steps that can mean green for you and your children.

It's the children who are our greenest ambassadors. Aware in ways we've never had to be of the need to conserve precious resources, our children will inhabit a planet in coming years that will be richer because of lessons taught by the Sally Foxes and so many others. The busiest environmental bodies in many neighborhoods are the young people, who frown if you don't recycle and pick up stray soda cans from the side of the road. A series like *Green Means,* and this book which sprouts from it, gives our children models and lessons which will benefit us all.

So, try to read the book and watch the series with an available child. Plant early the notions of what green means. And if you ever get to Washington, D.C.—our leafy federal city—keep your eyes open. If you spot a Clean Air Cab on the street, gliding along on non-polluting natural gas, hail it and tell Tayo I sent you.

GREEN MEANS
AN INTRODUCTION
Peter L. Stein, Executive Producer

The forty short television stories known as *Green Means* started out as a way of counteracting the general hopelessness that many people feel when confronted with yet another television documentary about the dire state of the environment. We in the media are partly to blame, of course, for this common but dour reaction; since the first Earth Day in 1970, many journalists and producers, in the course of raising the alarm about the greenhouse effect, ozone holes, and rainforest destruction, have often given the impression that the battle to save the earth has already been lost— and that the human impact on the environment is invariably a negative one.

But as grim as much of the environmental news is (and justifiably so), one message was getting lost: people *are* making positive contributions to the health of the environment—ordinary people, grassroots organizations, new technologies, unexpected heroes—all pointing the way to a healthier environment, and inspiring the rest of us to live gently on the earth.

And so *Green Means* was born. The idea was not to present a catalog of eco-heroes all in one program, but rather to produce an ongoing series of short profiles that might show up anywhere in one's television viewing. The programming professionals call this "interstitial material," but the rest of us inelegantly call them "fillers." In this case, we hope they are meaningful fillers: short bursts of green inspiration with useful tips about how to do right by the environment, even in our own backyards.

We set ourselves certain parameters in choosing the subjects of our profiles. They are parameters peculiar to television (more specifically public television), and have resulted in a collection of remarkable stories, but not necessarily the same ones that would be chosen by, say, a print journalist, an international panel of environmentalists, or a philanthropic organization.

Our stories needed first of all to be stories: the *Green Means* profiles needed to be actual dramas—each with a beginning, middle, and, we hoped, an end or result. Second, each had to have a visual element, for the biggest possible impact as a short television spot (this criterion eliminated many wonderful eco-heroes whose work was primarily technological, political, or theoretical). Third, we wanted our stories to showcase a wide range of environmental issues, from energy to biodiversity to population, and to comprise a wide range of geographic locations and communities (within budgetary limits). Fourth, our stories needed to stay current for several years, since they were meant to be "evergreen" eco-fillers for PBS. And finally, hardest of all, we needed to choose stories that could be told in about four minutes, without sounding hopelessly glib or superficial.

The result is by no means an exhaustive treatment of innovative environmental achievement, but I hope it's representative of what I see as a growing trend toward solution-oriented activism. Over the course of producing the first two seasons of *Green Means*, I've noticed a distinct shift in environmental discourse: one is now just as likely to read in eco-journals about restoration ecology, bioremediation, and international eco-collaborations as one is to find out about environmental disasters. I can only hope the mainstream media take up the trend as well.

Looking back over our first two seasons, a few production highlights come to mind. Actually, "highlights" is probably the wrong word; "nightmares" comes a little closer. Perhaps I am taking hardship too personally, but one would think that a series that was supposed to help the environment would have gotten a little help from the gods of earth and sky. Instead, they hurled their wrath at us. I suppose the problem was timing: we aimed to complete the first season by Earth Day (April 22) 1993, so shooting during the preceding winter was inevitable. But little did we know, we had picked the Worst Winter in Human Memory.

The litany of Mother Nature's rages to which our crews were subjected borders on the biblical. It started quietly: a steady rainfall dampened our shoot in Arcata, California, where we visited a series of manmade marshes that acts as the town's sewage treatment facility and doubles as a wildlife refuge. That was all right—everyone expects rain up in Arcata, and it was still very pretty. But then a freak December hailstorm hit sunny Santa Monica the day before we were to tape a group of eco-surfers; we learned that surfers don't hang ten in a hailstorm. The re-scheduled shoot, some six weeks later, coincided with pea-soup fog—now we couldn't even see the waves.

Then we went to Wyoming to profile Jack Turnell, a cattle rancher who has been dubbed the "Green Cowboy." The High Plains were spectacular in the crisp winter air—except for a sudden snow shower that blanketed the herds and sent the temperature plummeting to ten degrees below zero.

But the worst came as we prepared to tape our host, National Public Radio's Susan Stamberg, doing all of the introductions and closing tags in Washington, D.C. We chose a weekend just prior to spring's awakening, when buds on the cherry trees would be visible. Enter the "Storm of the Century," which kept the producer trapped and forced snowbound Ms. Stamberg to shovel out her driveway just to make it to the recording studio, where we completed the voice-overs by fax and phone.

Smarting from this experience, we planned to shoot the second season's stories in spring and summer 1994—a wise move, except for one small miscalculation: again, Ms. Stamberg bore the meteorological brunt. We shot her segments from dawn to dark under lights on a single blistering, muggy July day in Washington. At one point I think I heard her mumbling the word "inhuman."

But we weathered what the environment tossed our way, and the host is still talking to me. What we talk about often has the tone of surprised gratitude for what we've learned working with the people you will meet in this book: *Green Means* sheds a bit of well-deserved limelight on a handful of individuals, projects, and organizations who are doing extraordinary things for the earth's sake. In so doing, I hope it also inspires viewers and readers to take these people as examples of just how powerful one person's actions can be.

GREEN COWBOY

Jack Turnell, Pitchfork Ranch, Wyoming

*J*ACK TURNELL MANAGES one of the largest ranches in America, and much of his livestock grazing occurs on U.S. Forest Service lands. Although in many areas environmentalists have charged that public range lands are being overgrazed, their streams eroded, and water supplies depleted, Turnell's ranch is considered a model of environmentally sensitive ranching. A dozen years ago, though, like most ranchers, Jack Turnell didn't know the terminology of even the most basic ecological concepts. "I'd never heard the word 'riparian,' even though I went to college."

When grasslands are overgrazed, the original mix of plant species changes, altering an ecosystem that is particularly delicate in the arid and semi-arid climate of the West and Southwest. Eroded soil runs off into watersheds, clogging them, and suffocating many small forms of life. Streams can also be damaged by cattle congregating along the banks; trampling the soft, wet soil into the water makes the stream wider, shallower, and therefore warmer. The temperature change harms the organisms, including fish eggs, that normally live there.

"However, with proper grazing, this does not occur," says Turnell, who consulted with range and wildlife scientists to improve his range land and riparian areas—as he learned the streamsides are called.

Located in Meeteetse, Wyoming, the Pitchfork Ranch was founded in 1878 and claims a colorful historical note—Butch Cassidy committed his first crime, a horse theft, here. He did his drinking at a nearby saloon that is still in business.

The Pitchfork Ranch, one hundred eighty square miles of it, was bought in 1903 by the family of Turnell's wife. It has remained in their hands ever since. When Turnell took over managing it, in 1971, the spread was run as a traditional family ranch and still is. They punch cows pretty much the same way their predecessors did, with cowpokes on horses. The cattle are branded with old-fashioned branding irons, in an annual event that becomes a big social occasion with many neighbors and friends coming to help or just watch.

The ranch may be traditional, but its methods are progressive. Turnell raises some three thousand head of cattle on one hundred twenty thousand acres, while practicing sound environmental techniques and making a hefty profit, too.

The ranchland, ideal for raising cattle, includes thousands of acres of hay meadows and pastureland rolling across a valley floor down the middle of which meanders the Grey Bull River, lined with cottonwood trees. The ridge of the Absaroka Range shelters two sides of the Pitchfork Ranch from winter storms. Winters are mild as a result, with little snow. In the summer, deer and antelope roam here, along with herds of elk. According to Turnell, there are 2,000 elk, 1,000 antelope, and 800 deer, as well as bobcats, grizzly bears, black bears, mountain lions, coyotes, beavers, and foxes. "You might not see them at this moment, but they're here and we enjoy them. I can look out my window every morning and see those kinds of animals around."

The abundant wildlife lives in harmony with Turnell's cattle. Pointing to an antelope browse (the shrubs they browse on), he says,

"If you didn't ever graze this with cattle, the grass would crowd out the browse, and then eventually it would hurt the antelope herd. So it's important that you graze it a certain amount, in order to keep a balance between the grasses and the browse. All wildlife doesn't eat grass like everybody thinks. So if you want to have a healthy wildlife population, you have to understand those things."

The Pitchfork Ranch land, including a river, a large creek, and many small creeks, is prime riparian area. The definition of riparian, for those of us like Turnell who didn't learn it in school, is, according to the Riparian Association, "the green area immediately adjacent to such surface water features as streams, springs, rivers, ponds, and lakes. This area is identified by vegetation that requires unbound water in quantities greater than that which falls on the area as precipitation. (While the terms wetlands and riparian habitat often are confused, wetlands is a more general term that also includes bogs, marshes, swamps, prairie potholes, and riparian corridors.)"

To protect the Pitchfork's riparian areas, Turnell installed watering tanks high on the hills to encourage the cattle, as well as the wild elk, antelope, and deer, to move away from the lower streamlands. To further protect riparian areas, he began systematically rotating cattle-grazing pastures, to keep cattle out of riparian pastures until after the season when the streamside grasses matured and spread their seed. As the streambanks improved, the trees grew and erosion subsided. He gave up most fertilizers and pesticides to help improve water quality; he finds he doesn't need them. To increase duck habitat, he teamed up with wildlife groups.

Becoming a green cowboy involves "a million things," Turnell says. "Little things, but they protect the resource. Like how you move cattle and how many you move at a time." One of his changes was to raise fewer Hereford cattle and more Salers, a breed from south-central France. Previously, with the Herefords, Turnell says, "we used to go up there and see 1,100 to 1,200 cattle, most of them on the creek beds for most of the day. Now, the Salers range two miles off the creeks. They aren't lying around the creek bottoms and beating the hell out of it. With the Salers, instead of taking one thousand head and pushing them up the side of the mountain like we used to, we just take a couple of hundred up each day and put them into a holding pasture and they drift up, scattering themselves out. They just go and go, out in the boonies."

The progress of the changes at the Pitchfork Ranch was monitored by photography, which showed the damaged riparian areas rapidly healing and providing better forage, while the water quality and fisheries also improved. With better forage, Turnell began making more money because the better quality grass puts more meat on the cattle. "We're producing 542,000 more pounds of beef per year than we did in 1987," says Turnell, whose ranch productivity ranks in the top few percent in the U.S. An average ranch raises three hundred head of cattle, or less. At the Pitchfork, "we've increased the pounds on the cattle and improved the range at the same time. At an average of 80 cents per pound, we've earned $433,600 a year, without any extra cost of raising them. And I can see down the road, it's just going to get better." By comparison, Turnell says, "If a ranch is grazed at the wrong times for the wrong reasons, after a period of years you're going to end up destroying the habitat and the grasses. And if you don't have enough grass, gradually the pounds of beef go down and down and down, and it doesn't take a rocket scientist to figure out that's going to lose you some money. You just convert the grass to pounds of meat and the pounds to money. We improved economics, improved cattle, improved wildlife, and it started making sense to me that you do all these things together."

Until a few years ago, Turnell wouldn't listen to environmental groups or even his own family when it came to running the ranch that's been in his wife's family for six generations. "I suppose I was quite dictatorial, and I didn't need anybody's help. I had all the answers. And it's been hard for me to change and to say I don't have all the answers."

Turnell began to change after a day in 1981 when his neighbors found the dead body of a little furry animal they had never seen before. They took the dead body to the local taxidermist to have it mounted. But the taxidermist told them that the creature was a black-footed ferret, which was listed as an endangered species and could not be mounted.

Some authorities thought the black-footed ferret had already passed into extinction. Ferrets had once thrived in this region, hunting the plentiful prairie dogs of the Great American Plains that stretched from Texas to Canada. As towns and cities began springing up on the Plains, prairie-dog towns disappeared, plowed under or cemented over. As the prairie turned into housing developments, the remaining prairie dogs became known as backyard pests. Huge numbers of them were killed off. When the prairie dogs got scarce, the ferrets soon disappeared, too.

Turnell's neighbors reported their find to the Wyoming Game and Fish Department, and a small colony numbering 18 black-footed ferrets was found on the Pitchfork Ranch and surrounding areas.

"Then I started receiving calls, and everybody in the world was excited because we had these black-footed ferrets," says Turnell.

Most ranchers live in fear of a situation like this: an endangered species is found on their land, soon followed by strict governmental, bureaucratic regulations. In fact, many ranchers refuse to cooperate with scientists—even for simple land-use studies—because the ranchers say they're afraid that someone will find an endangered mouse, or insect, or something, and ruin their livelihood in efforts to protect it.

In Turnell's case, sure enough, a stream of scientists and environmentalists poured over the Pitchfork Ranch. "The ferrets forced me to tolerate people who I'd traditionally been an adversary of," says Turnell. "I found out, by God, they were people, and they were interested in something good." The stream of people visiting the ferrets turned into a flood. Round the clock they came: scientists, reporters, authors, filmmakers, and wildlife lovers. Turnell welcomed them and, later, when he concluded that the general public did not understand the importance of the animals, he helped finance a public television documentary on the ferrets. On further investigation, the original colony of 18 ferrets was found to number more like 130. Turnell joined a ferret advisory committee whose goal was to bring the little animal back from the brink of extinction.

Gradually, Turnell stopped fighting with the environmentalists and the bureaucrats. Now, he says, "I work with them, not against them." He started making changes in his ranching and moved his cattle to protect the ferrets; he modified oil exploration and drilling in the ferret study area.

After his involvement in saving the ferrets, Turnell opened up to other opinions. "Well, it just simply caused me to cooperate with people. And all of a sudden, in the end, after dealing with lots of people and lots of interests, and trying to help save an endangered species, it got me to thinking that maybe I could work with a lot of people to save the other things, the riparian, the range, the wildlife as a whole. And so I guess I credit the ferret discovery with a lot of my views today and what I'm going to do tomorrow, or down the road. It had a big impact."

Converted, Turnell is now committed. He helped form the Wyoming Riparian Association in 1989, for livestock producers, water management agencies, and environmental groups. The goals of the association, with Turnell as chairman, include education and cooperation.

In many other ways as well, Turnell works "to close the gap between reasonable environmental groups and reasonable farmers and ranchers." His ranch hosts tours for the U.S. Forest Service and other groups to show that environmentalism and profitable ranching practices can go hand-in-glove. He's got a project going with the local school to give students a chance to come to the ranch and work on environmental issues. On the other side of the fence, he gives talks about conservation to groups of cattlemen.

"My whole life's goal is to convince the world that we can do all of these things—we can drill for oil and not disturb the deer and elk populations; we can graze cattle on the public lands and actually improve the range conditions for all of the wildlife." He calls his philosophy "cowboy science" and each year he spends more than $10,000 out of his own pocket promoting it, to show that environmental concerns and profitable ranching can coexist.

Getting other ranchers to agree with him has been an interesting task. At first, he says, they thought he'd lost his marbles. "The truth is, five to eight years ago I had a lot of flack, but now they're starting to understand what I'm saying. It's not because I know everything about range or watershed management. I don't. They're understanding that the key to our survival is to communicate. You know, I'm not a psychologist, but I'd like to believe that I understand both sides of the issue. If cattle ranchers and environmentalists both can change," he says, the future looks bright. For his work, he has won a handful of environmental prizes regionally and nationally, and appeared on national television shows, including *ABC World News Tonight.*

Now, he says, "the livestock industry should complete the century by taking the lead in declaring that they are environmentalists, conservationists, and managers of wildlife and Mother Nature's resources—then carry out plans that make it a reality."

FOR MORE INFORMATION:

Jack Turnell, Pitchfork Ranch, Meeteetse, WY 82433.

WHAT YOU CAN DO:

You may not own a huge ranch, but even in your backyard you can practice good land management. Let your lawn grow shaggy—at least two or three inches high. Your grass will dry out more slowly and need less watering.

When you mow, don't throw out those clippings. Compost the clippings, along with other yard cuttings, and you can make your own organic fertilizer. Most lawn mowers can be fitted with an attachment that automatically mulches the grass clippings and spreads them back on the lawn. The attachment costs as little as ten or twenty dollars. Yet, currently in California, some ten million dollars a year is spent sending grass clippings to landfill dumps, according to the California Integrated Waste Management Board. New additions of grass clippings in California alone total approximately 3.4 million tons—enough to fill a football field piled 306 stories high.

Nationally, yard waste is estimated to account for nearly twenty percent of the municipal solid waste generated each year. Grass clippings make up more than half of the yard trimmings. So, don't bag it—mulch it or compost it.

THE BUFFALO RETURN

Fred DuBray, Cheyenne Sioux Reservation, South Dakota

*F*OR HUNDREDS OF YEARS, the indigenous people of the Plains centered their lives, culture, and spirituality on the buffalo, seemingly designed by the hand of nature for the purpose of supplying nearly all of their needs. From the buffalo they made food, clothing, shelter, boats, shields, ropes, bags, ornaments, eating utensils, and drinking vessels. Even the dried dung served as fuel. "It was like a commissary," says Fred DuBray, president of the InterTribal Bison Cooperative, dedicated to bringing the buffalo back.

The Indians and the buffalo were cleared from the American West together, like two pests in the white man's promised land. Soldiers, railroad builders, miners, ranchers, trappers, and settlers quickly saw that as long as the buffalo provided all the Indians' needs, the Indians could live freely and independently. Millions of buffalo were slaughtered as a strategy to defeat and extirpate the Indians. And sure enough, without the buffalo, the Plains Indians lost their strength and courage, and their land was taken. The tribes were betrayed, broken up, scattered, and thrust onto reservation lands where most have lived in poverty and despair for the past one hundred years or more.

Even if the Indians had not been in the way, the buffalo would have been. Vast herds of the huge, fierce beasts dominated grasslands that ranchers wanted for docile cattle, and grazed where settlers wanted to plant crops. The beasts had to go, and they did.

Hunters routinely killed five times as many buffalo as they used. The "distinguished Buffalo Bill Cody" is credited with a record kill of 4,280 buffalo in 18 months, 48 of them in 50 minutes. The hunters used only a small percentage of the carcasses, and of those, they frequently cut out only the choicest morsels, sometimes just the particularly delectable tongue. The rest of the animal they left to the wolves and vultures.

When Europeans came onto the Plains in the 1840s, there were an estimated forty-to-fifty million buffalo. A mere fifty years later there were only one thousand. The buffalo would have become extinct except for a few farsighted men, both Indian and white, who captured some of the last wild buffalo in the 1880s and 1890s. This is the only reason we have any of them left to us at all.

One hundred years later, in 1990, the InterTribal Bison Cooperative (ITBC) that DuBray heads was formed to reintroduce buffalo onto the reservations. An organization of twenty-eight tribes from thirteen states and one Canadian province, they believe reestablishing healthy buffalo populations on tribal lands will help reestablish hope, pride, and prosperity for Indian people. "We recognize the buffalo as a symbol of our strength and unity, and that as we bring our herds back to health we will also bring our people back to health," says DuBray.

"In Indian spirituality, everything is related. As Indian people have developed and evolved throughout the ages, they have done so alongside all of these other creatures—buffalo, prairie dogs, eagles— all these different things that they learned from. They recognized the importance of all these other things and that the people were only part of this grand scheme, not any better, not any worse, but just a major part of it. So one of the terms that they used to describe that is translated as 'all my relatives.' And that's what the spiritual essence of Indian people is, that connection to the rest of the world. The Indian spirituality is not the same as an organized religion like Christianity or something like that. There is no such thing as Indian religion. It's a way of life that people have. It's a way that they look at the world and live in harmony with it. So although they were constantly changing and evolving with the world, it was a real gradual process because they were in tune with it, in harmony with it.

"You have to look out for all these other things as well, not just yourself and your own kind, because you're dependent on them as well. Indian people call all these things their relatives because in our own social structure, your relatives are real important, and you have to take care of them. If you don't have any relatives, then there's nobody to take care of you. If you're related to everything, then everything will take care of you, everybody will take care of you. So you keep it that way, and then you're also looking out for your own welfare. We put the spiritual aspect of these buffalo in relationship with the people at the forefront rather than economics, because we firmly believe that if we can keep the spirituality alive the economics and politics will follow suit."

To date, the ITBC has a collective herd of about five thousand animals on tribal land from Maine to California. The number has increased from two thousand animals in 1990, when the nonprofit organization was formed. In ten years, they estimate the herd will reach twenty thousand. The tribes include the Blackfeet, Cheyenne River Sioux, Choctaw, Crow, Nez Perce, and many others.

DuBray is Lakota, and he says this is what he prefers to be called, rather than Indian or native American, which he says are "white man's terms that have been placed on us. The reason we became Indians is because Columbus was lost when he came over here. He thought he was in India, so he called all the people here Indians. The term stuck. Even 'America' is a European term for this country. We didn't have a name for the whole country because we each had our own little niche within it. So when they call us native Americans— well, you have people who were born here that claim to be native Americans regardless of what race they are. I've heard a lot of people say, 'I'm a native American because I've had three generations of ancestors born here. That makes me a native American.' So it's not an accurate term either. The only accurate term that each one of our tribes has is their own name, which in our case is Lakota." The people in general, though, he calls Indians.

The ITBC holds to a "strategy to keep the buffalo and market them as strictly native and natural. When you start into the domestication process, you start changing the animal somewhat. They are big, dangerous animals, and you find yourself eliminating the more aggressive ones for your own self-preservation." Currently, apart from the ITBC herds, there are more than 130,000 buffalo being raised on more than one thousand ranches, most of which are little more than converted cattle feedlot operations. Buffalo farms are managed by breaking up social units of the animals, dehorning them, and using intensive management techniques such as artificial insemination, constant applications of drugs, and premature slaughter. The proud spirit of the buffalo disappears, and an animal much more like a cow results.

On the lush, bluestem grass prairies, herds of Plains buffalo moved in an annual migration north in spring and south in late fall, as far as four hundred miles round-trip. The routes rarely varied, year after year, and some trails were worn three feet deep. During the journey, the huge procession of great, shaggy beasts stopped or even derailed trains in their path. They slid down steep banks and cliffs and swam swift rivers, sometimes halting boats in their way.

The buffalo evolved the landscape, as they evolved themselves. Always on the move, the herds grazed on grass as they walked. The taller growth they left behind, and that provided shelter for ground-nesting birds, mice, lizards, and other small animals. "They churned the ground with their hooves as they passed over it in large numbers," says DuBray, "which actually did the ground some good. Then they moved on, and the land continued to grow. After the buffalo moved through, the prairie dogs had a heyday because they need help in keeping this grass graded down. Their only protection from predators is being able to see them approaching from a distance. So they graze around their holes and try to keep it chewed down so they can see when the coyotes come. Well, if the grass is too tall, they can't see, and their numbers don't do so well. They don't multiply very fast because predators keep them in control. So when the buffalo come there and eat all that grass off, well, that helps the prairie dog. The prairie dog population multiplies tremendously as they pass through. But then the buffalo move on. They don't hang around. The buffalo come back maybe a year later into that same area. By that time, the grass has grown back up again, predators have moved back in, and prairie dog numbers are about back to where they were." The buffalo also fertilize the soil with their droppings. In the spring when they shed their coats, seeds caught in the hair begin to put down roots in the soil and grow.

Although wolves might attempt to take a weak, sick, or young buffalo, on the whole, the mighty beast evolved without predators. The animal put its ferocious power into duels for dominance, where two bulls collide head and horn. During rutting season, April to July, a dominant male stays with a herd of ten to fifty cows, calves and young bulls. He is challenged over and over, and the slightest weakening of his strength sends him off to the sidelines. "Fighting storms" break out every few days, and within two hours, fifty or sixty fights may erupt. Every challenge is met with a charge. Relationships are rapidly rearranged, and, as a result, new genes are mixed into the breed and passed on. Because only the most ferocious bulls successfully breed, natural selection results in the perpetuation of a powerfully spirited animal.

The cows give birth from April to June, to a twenty-five- to forty-pound calf who is on its feet and suckling within thirty minutes. As another rutting season begins, the young bulls, just a few months old, exhibit the vigor of the breed, butting heads together in play. In five years, the bulls will be as large as their fathers and capable of bellowing out a challenge to more mature males for the rights to a cow. The younger animals, leaner and stronger, force the older bulls aside. The dominance battles ensure that only the strong thrive.

The buffalo and the Indians lived on the prairie in ecological balance for centuries. The Indians knew intuitively the basic rules of ecology. In the words of an elder Indian: "The killing of buffalo enables the Indian to live and grow, and when his mortal remains return to the Earth it serves as food for grasses of the prairie, which in turn feed the buffalo, thus ensuring the constant cycle of life."

Then the buffalo and Indians were replaced by cattle and Europeans. "The cattle just stay there and keep on grazing around these same areas," says DuBray. "So that kept the grass down, so the prairie dogs just kept on multiplying. That's caused by overgrazing. And that leads to poisoning the prairie dogs. It really doesn't need to be done, because all these animals are really quite capable of taking care of themselves. That's the ironic part of it. The more people try to deal with it, to manage and control all these things, the worse they mess it up."

Buffalo meat is leaner and contains more protein than beef and is considered very tasty. It was, after all, first hunted as food by Indians and white men alike. A market for the meat as a specialty food already exists, although it is expensive. Prices range from about five dollars a pound for ground buffalo in Boulder, Colorado, to thirty-five dollars a pound for prime buffalo ribeye steak from a mail-order firm. A full-length buffalo robe coat costs about two thousand dollars. Clothing and craft products from other parts of the animals, such as pillows, painted skulls, and horn jewlry, are sought worldwide.

Raising the buffalo for market makes good sense, but "if you just replace all the domestic livestock with buffalo," says DuBray, "it doesn't change anything but the color of the animal. It doesn't fix anything." Buffalo must be treated as wild animals rather than as domestic cattle. The buffalo are the Indians' partners in survival, not just an exploitable commodity.

The buffalo-raising project, DuBray says, "is a strategy that developed from grassroots ideas. It's something that's been around for thousands of years. It fits with the culture. It fits the spiritual, economic, environmental, and social needs. The Indian people don't separate these things out. They all go together, holistically."

To preserve the spirit as well as the physical characteristics of the American buffalo, the ITBC wants to let them remain wild, in their natural state. At the same time, the ITBC intends to harvest the massive animals and get them to market. The tribes are still searching for the best way to make the two goals meet. "If you put them into a corral," says DuBray, "they just go crazy trying to get out. All they can think of is getting away, and they do a lot of harm to themselves, not to mention the facilities. As you try to get them in closer confinement, they get worse. By the time they get to slaughter, they're just so freaked out the adrenaline and hormones are running through their bodies, and you lose the quality of meat.

That happens with deer or any other animal. If you run them around the country and shoot them up a few times, they're not going to taste very good."

DuBray works with a mobile facility that goes out into the pastures and allows harvesting the animals in their natural setting. "You don't upset them that much, and you reduce most of the problems that are associated with trying to harvest them the way that people are doing today."

The Cheyenne Sioux Reservation has a population of ten thousand people on land the size of the state of Connecticut, where they keep about nine hundred head of buffalo. They are confined, if that word can be used, on twenty thousand acres. "We've got certain areas that we fence off and try to keep them in. Sometimes you'd have to say they are free-roaming, because they kind of go where they want to go, when they want to go. But there was already a lot of cattle fence running around, so we just added a couple of wires to it to make it a little bit higher. We don't want to restrict the movement of other animals, like antelope and deer. Of course, it does allow for buffalo to move through, too. But there isn't much of a fence that you can put up to keep them confined anyway, unless you want to put them in prison-type situations, which is not what we want to do."

DuBray was born on the reservation and sent to boarding school. Later he attended Black Hills State University in South Dakota. "I went there because it is in the Black Hills, which are sacred to our people. I wanted to be in that environment." He took courses in social science, communications, political studies, and Indian studies. "Indian studies was really kind of a different experience." All of the history was written by white people, because the Indians had no written history of their own, and just about all of what he read was jarring.

DuBray sees the treatment of the Indians as an extension of the way people treat the environment. "Ever since people have taken over this land, they've wanted to change all of the plants and animals—and people—into a domestic type of species. Replace the wild and native stuff that was here with their own stuff. You take away all the wildlife animals that were native to the country and replace them with exotic animals. So then you've got to keep changing the environment to adapt to all these exotic animals that you put on it. In my opinion, the problem is the whole guiding philosophy that people are superior to all other species, somewhere above the animal world, kind of in between the animal world and God, and always striving towards being more like the gods. It causes a lot of destruction of all the other species, because they are looked at as inferior and therefore less important. And they are being eliminated on a daily basis. Species after species is becoming extinct. Which is a self-defeating philosophy because we need all of them to survive ourselves. Once they disappear in large enough numbers, then we're gone, too."

The Indians' struggle for dignity and independence is a struggle against forces in power, DuBray says. "But if you're the one that's in power, you have a struggle going on, too, only it's within. You don't look at it as the same type of struggle because you're the one on top." The ones on top feel they have to defend their position, "and with a certain sense of self-righteousness, they think that other people who are struggling are trying to upset them from their place, to take over their place. That's not true, in our case. The Indian people don't want to be in that place. We don't want to control anybody else. We just want to be able to control our own lives and be able to do our own thing."

FOR MORE INFORMATION:

InterTribal Bison Cooperative, 520 Kansas City Street, Suite 209, Rapid City, SD 57701.

WHAT YOU CAN DO:

Individuals and organizations interested in supporting the nonprofit ITBC may become honorary members for an annual fee of $100. All contributions are devoted to sustaining and strengthening Indian buffalo projects. Honorary members receive a quarterly newsletter and periodic technical publications on the cultural, social, ecological, and economic aspects of buffalo and their relationship with the Indian people. To make a contribution, contact ITBC at the address above.

DuBray also suggests you remember the spirituality of nature and the environment. You don't have to have three million acres of land. "You can look at one square foot of ground, and if you study that long enough, you'll see a lot of activity taking place there, constantly. It's that kind of connection you have to start with and understand."

CRIMES AGAINST NATURE

Ken Goddard, Ashland, Oregon

*M*OST CRIMES AGAINST wildlife leave only bits and pieces of the animal victim's body behind. Sometimes investigators can't even identify what animal the pieces came from; the evidence might be only a pile of guts in the woods. The new National Fish and Wildlife Forensics Laboratory can now positively link animal parts, such as a gut pile in the woods, with other evidence of a suspected crime, such as a trophy head. Further, the lab can identify the same animal's blood found on a knife or even under a suspect's fingernail. Using physical evidence, the forensics lab can link the crime scene, evidence, victim, and suspect. Previously, a criminal typically had to be caught in the act of killing an animal or selling animal parts or products to be convicted.

"Now we have the capability of putting scientific fact before the jury—just as is done in police homicide investigations—and let the jury decide what happened," says Ken Goddard, director of the lab. "We can match the meat in the freezer, blood on a car, and blood on clothing with one animal and no other animal in the world with absolute certainty." The wildlife forensics lab peers into DNA, the genetic material of all life, found in all blood, skin, and hair. DNA creates an individual's specific characteristics. The procedure of identifying specific individuals from DNA is commonly called "DNA fingerprinting."

The wildlife forensics lab has also made another breakthrough—in identifying not only the victims, but the actual bullet that caused the death. Crime scientists already knew that the fine lines and nicks left on a bullet as it shoots out of a gun barrel are unique to that weapon. But matching the guns and bullets was a process that required several hours per bullet, until one of the scientists in the wildlife forensics lab "tripped across" a mechanism for laser scanning and digitizing that allows high magnification and computer comparison. The information can be entered into a database, much like fingerprints are now, and the computer can look at several thousand bullets, or even several hundred thousand, in a very short time. The bullet's characteristics can be traced even if the bullet is damaged by impact with an animal. "This has never been possible before," says Goddard. "It is really an astounding discovery. It's too big for us. We turned the procedure over to the FBI for development with the hope that they can use it for some sort of a national database."

Compared to the FBI, the lab staff is tiny: thirty-two people, of whom eighteen are scientists and the rest clerical and support staff. Their facility is large, though—some twenty-three thousand square feet, tucked out of the way on the campus of Southern Oregon College in Ashland.

Born and raised in San Diego, Goddard received his undergraduate degree in biochemistry from the University of California, Riverside, in 1968, and joined the county sheriff's department. "By day they taught me forensics, and by night they took me out on patrol. I loved it." He went on to get a master's degree in criminalistics from Los Angeles State University, while working in crime labs in Southern California. "You start out idealistic. You rapidly discover that you have virtually zero impact on the flow of rapes, robberies, murder, drunk driving, dope. The numbers keep coming. You get a real sense of shoveling against the tide. As humans, looking at human crime, we become fully callused."

Goddard found the police work progressively depressing. "The way you maintained your sanity was basically to develop your blinders. You learned not to react to the bodies." Once, at a homicide scene where a man had been shot, after hours of studying extremely confusing evidence, Goddard and his fellow investigators sat down on the bed to puzzle over their notes. "One of us looked down and realized that all three of us had our feet up on the body as a foot rest. The body had long since become a piece of furniture. And that's how you get. It's a necessary defense mechanism. I don't think that's particularly healthy for you on a long-term basis."

In 1979, burned out on human crime, he answered a help-wanted ad for the first chief of forensics for the Fish and Wildlife Service. Once he had the job, he found out the Service didn't have a lab or any budget for one. "It never occurred to me that they would hire a police crime laboratory director to set up a forensics program and not have plans to build a lab. So I had never asked about it." While lobbying for a lab, Goddard went out into the field and worked on catching criminals by collecting low-tech evidence such as fingerprints, footprints, and tire tracks. But wildlife crooks are getting more sophisticated all the time, with automatic rifles, organized gangs, and trade routes. They are motivated by an estimated two to three billion dollars in illegal wildlife trade; the head of an elk with large antlers can sell on the black market for twenty thousand dollars.

In an effort to keep up with the outlaws, the Fish and Wildlife Service beefed up its division of law enforcement, beginning in the mid-1970s. Hiring Goddard was one of its steps. Another was to send their new federal agents to intensive FBI-type training in techniques of criminal investigations, including knowledge of firearms, self-defense, and wildlife law.

Goddard and his crew in white lab coats investigate cases brought to them by wildlife law enforcement agencies around the world. A fairly typical case involves big-game hunting of an animal protected by law, or hunting during the off-season. The murder method varies: the animal may have been stabbed, poisoned, or shot by bullets or arrows. The method may be more subtle, as in one case when a tiger was drugged and then killed without an external mark: an ice pick was pierced through the ear canal into the brain. In another case, the method was deliberately obvious: a rare spotted owl was found dead, nailed to a tree. The spotted owl has held up timber sales in the Pacific Northwest so that its habitat can be protected.

Another category of cases involves illegal animal parts and products, such as crocodile handbags, rattlesnake boots, leopard-skin coats, ivory necklaces, bear-paw ashtrays, eagle-beak paperweights, and even sea turtle mandolins. Trophies show up, too: hippo skulls and python skins. Then there are the elixirs, potions, pills, balms, and tonics made from rhino fetuses, sea turtle eggs, elk horns, tiger whiskers, or seal penises. The seal penis is claimed to be a male aphrodisiac, as are many other animal parts, including rhino horn and tiger bone. Goddard calls it "the guy problem," because this particular market demand alone is driving some species, such as the black rhinoceros, toward extinction. "It's a sad commentary on us guys. The idea is that if things are not going well for one of us, we would grind up a dried seal penis into a powder and take it as a tea, thereby supposedly making things work out better. I not only remain skeptical, I really don't think I could handle the cure. I'd have to stick with the problem." He has in his facility a big sack of what are purported to be seal penises. "But what we're finding is that many of the things that are sold as seal penises turn out to be canine. You know, there are lots of dogs that go through the dog shelter euthanasia program. Presumably, in some locations, parts are collected."

Aphrodisiacs are only one kind of remedy; Oriental traditions prescribe animal parts for a long list of ailments, from calming fright to curing cancer. "I think one of the more interesting discoveries we made in the laboratory was that the Asian medicinals involving endangered species tend to be fake. Virtually every rhino horn preparation we examined in the laboratory—and we've examined hundreds—have all been fake. There's no rhino horn in there. In fact, typically they contain large amounts of mercury and arsenic. The irony is that instead of being a supposed aphrodisiac with health-restoring capabilities, what we have are placebos that contain heavy-metal poisons. The same has been true for the tiger bone preparations and many of the other Asian medicinals we've examined. Pretty much everything turns out to be fake. Yet, at the same time, we're constantly finding the rhino horns, tiger bones, and pieces and parts of the endangered species animals. The sense we have is that the vast majority of the market is fake, but wealthy individuals have access to the real things."

A third category of cases is poisoning. "Let's say a rancher believes that eagles are going after his sheep—which eagles typically do not. They are scavengers. They don't kill large animals typically, but if they see a dead lamb down there, they're going to land on it as a food source, and if the rancher happens to see an eagle on top of a lamb, he's probably going to assume that the eagle killed it. So what's happening is that some of these folks are going out and dosing carcasses with huge amounts of very lethal poisons. Then an eagle lands on it, eats, and dies. Something lands on the eagle, eats, and dies. You have secondary and tertiary kills going on in all directions. We investigate those. Obviously, we have to be extremely careful in handling the poisoned carcasses, because whatever shut off that eagle can certainly shut us off, too."

Those are the main categories of cases, but "day to day, one never knows what to expect," says Goddard. "Anything can arrive and does—usually unannounced." Or a crew might be sent out to the field, as they were when a fifteen-hundred-pound walrus carcass, with the head missing, turned up in Alaska. Some odd requests come in, such as to investigate samples that supposedly came from Bigfoot, the mythical, half-man, half-beast of the Pacific Northwest. "I basically told them to go away as politely as I could. I used the somewhat flimsy excuse that this wasn't a law enforcement issue because it's not illegal to kill Bigfoot." Then someone managed to have it declared illegal to kill Bigfoot and the lab agreed to examine the samples. "You may be one of the first to know," says Goddard with a laugh, "that Bigfoot seems to have polyester fur—a fascinating adaptation."

In the few years since the scientists settled down to their microscopes and spectrometers, chromatographs and spectrophotometers—a million dollars worth of sophisticated equipment—they have led the whole science of wildlife forensics forward by several quantum leaps. The DNA breakthrough and bullet digitizing are two such leaps, and actually not the earliest. In the first six months of operation, the lab discovered a way to distinguish illegal ivory from legal ivory. The legal stuff comes from prehistoric mastodons and mammoths, at least thirteen million pounds of which is reported to be in the deep freezer of the Arctic tundra. If you unearth those tusks, you're allowed to keep and use them any way you desire. But elephant ivory has been illegal since 1989.

Suddenly, agents started seeing a big increase in "mammoth" ivory shipments. "There was no way to verify that ivory marked 'ancient' was really legal mammoth or mastodon ivory and not mislabeled illegal elephant ivory," says Goddard. There also suddenly appeared many ivories labeled "warthog" or "hippo." In 1990, using a $250,000 microscope and a 25-cent protractor, the scientists discovered structural differences between all the ivories, whether ancient or modern, warthog or hippo, walrus, narwhal, or sperm whale teeth. Not only that, they learned how to identify powdered rhino horn as well.

Another breakthrough occurred in the black bear gall bladder case. That case revolves around the fact that ancient Asian medicine prescribes processed bear gall bladder for pain relief, as well as to treat bruises, abscesses, cataracts, gallstones, and cancer. In Asia, a bear gall bladder sells for as much as a thousand dollars a gram, and as a result the Asian bear population has nearly vanished. Poachers have turned to North America, where they illegally kill roughly three thousand black bears a year, leaving the carcass behind; only the small, pear-shaped organ, and perhaps the paws, are removed. The bear gall bladder can be shipped and sold under the label of another, legal animal, because whether it comes from a pig, a bear, or any of many other species, gall bladders of many species look alike—especially once they have been dried or pulverized. The lab's sleuths discovered, however, unique bile acids secreted by the gall bladders of black bears and can now positively identify the organ. Their continuing research revealed that roughly eighty percent of animal gall bladders sold in Asian marketplaces from the United States come from pigs.

As the lab scientists are able to scrutinize ever more closely the foundations of animal life, science slides into philosophy. One such slippery slide occurred with the case of the Yellowstone wolf. The animal was found shot near Yellowstone National Park, where wolves are protected under the Endangered Species Act. Before starting to prosecute anyone, though, the dead animal was sent to the wildlife forensics lab to determine whether it was actually a pure wolf, because hybrids are not protected under the law. "So we cheerfully went out to find a known, pure, one-hundred-percent standard wolf against which we could compare the DNA. We got brought up rather short. Where could we find such a thing? A pure, one-hundred-percent wolf that's never had a sled dog or domestic dog in its heredity ten, twenty, fifty, or one hundred generations back? So we went looking for the genetic definition of a wolf and discovered there is no such thing. Basically, if it looked like a wolf and acted like a wolf, for all practical purposes, it was a wolf. Well, that's all very fine for a wildlife manager, or for the wolf for that matter, but it's not so fine if you have a lawyer in court saying, 'My client killed a hybrid. I dare you to prove differently.'

"There was a great deal of interest in the case, and everybody was waiting for the evidence. They were calling us up on a daily basis asking how soon the DNA would be done. And we were still looking at each other asking, 'What the hell do we do? Where do we find our definition, not to mention our standard?'" The lab crew tried several routes of scientific inquiry, and "after about three months of this, everybody was getting pretty well fed up with us. But in the process of thrashing around, we got some samples from a place called Nine Mile Valley, about three hundred miles northwest of Yellowstone, which has a population recognized by the Fish and Wildlife Service as being genuine wolves. It turned out to be an exact DNA overlay. So we resolved the case by writing a report saying this is a wolf by definition since the matching creatures are defined as wolves." He pauses to laugh. "Of course, later it occurred to us to wonder, 'Well, wait a minute—how would the Fish and Wildlife Service know those creatures are pure wolves?'"

Interesting as philosophical questions are, though, they are not the role of a cop or forensic scientist, Goddard says. "We present our evidence, and we go away. We have an absolute obligation to be unbiased and ethical in our work because we are helping to put people in jail and taking away their livelihood, based on our scientific results. There's a tremendous pressure on us in the face of the courts and attorneys and cross examinations and all that sort of thing. But there is tremendous satisfaction also. We are basically trying to give the animals a fighting chance. Give them a chance to maintain themselves in their habitat. Humans cheat; they violate laws; they try to take more than their fair share. It's good old human greed, and that's what our laboratory is basically here to fight. I think that all of us as law enforcement officers and officials realize that law enforcement does not resolve issues. Education is the key. Education is what's going to resolve the wildlife issues. The public has to become involved in not purchasing the parts and products and become involved in the habitat issues and the preservation issues and finding that balance between people and animals and habitat."

FOR MORE INFORMATION:

Ken Goddard, Director, National Fish and Wildlife Forensics Laboratory, 1490 East Main Street, Ashland, OR 97520.

WHAT YOU CAN DO:

The United States is the world's largest wildlife-consuming country. Of course you wouldn't knowingly buy any product containing parts of an endangered animal—but how do you know for certain what those are? The list, unfortunately, is very, very long and the number of crooked dealers is increasing, too. The wise consumer will beware of buying any products made from animals, especially when traveling internationally. The United States is one of the 122 countries abiding by the Convention on International Trade in Endangered Species (CITES) Treaty, basically agreeing to enforce each others' endangered species laws. So even if souvenir items are sold openly in another country, if they are illegal, you could be stopped at U.S. customs. You will, at the least, be relieved of your expensive souvenirs, and, you might possibly face a fine as well. According to agents of the Fish and Wildlife Service, the most common illegal wildlife product that people bring back as a souvenir is some part or all of a sea turtle, whose population is in danger worldwide. Common sea turtle products that are illegal to bring into the United States include lacquered turtle shells, shell jewelry, sea turtle soup and face creams, to name a few.

For more information, write for the brochure titled "Buyer Beware!" from World Wildlife Fund, 1250 24th Street NW, Washington, DC 20037. Another helpful brochure is titled "Facts About Federal Wildlife Laws" and is available from your nearest Fish and Wildlife Service Office, or write to Division of Law Enforcement, U.S. Fish and Wildlife Service, P.O. Box 3247, Arlington, VA 22203-3247.

REEF RELIEF

DeeVon Quirolo, Key West, Florida

*C*ORAL REEFS—mysterious, fascinating, and teeming with life—are built by simple little animals over thousands, even millions, of years. People plunge into the world of the reef, crashing through its sunlit ceiling into an underwater garden of color. Rocks and fish flash iridescent in shifting sunbeams. Silence falls deeper with the diver's descent, and peacefulness prevails. Yet a tension hovers between predator and prey. Sharks and barracuda stir panic with the flick of a tail; with another flick, tranquillity is restored. The well-balanced ecosystem thrives by allowing each animal, from the coral polyp no bigger than the head of a pin to the eighteen-foot shark, its niche. The reefs have proved resilient through the ages, surviving ferocious hurricanes and building again. Now, though, after 450 million years on Earth, they may have met their match: human beings.

Scuba diving and snorkeling have grown in popularity so quickly that too many people with too little understanding are destroying too much. The coral reefs are crumbling and vanishing. The causes are readily apparent: people are dropping anchors on reefs, standing on coral, dragging equipment, harvesting live rock, overharvesting marine life, stirring up sediment, dredging with boat propellers, and disturbing nesting birds. When you add pollution, intensive agriculture, coastal development, and global warming, the course is set for devastation. There is very little time left to turn the foolish ship around.

An annual wave of four million visitors washes over the Florida barrier reef. Just six miles off shore, the reef is some 158 miles long. It begins south of Miami and parallels the string of tiny islands called the Florida Keys all the way down through the city of Key West and out past the Dry Tortugas. It is North America's only living coral barrier reef, and it is the most visited reef in the world. In the Florida Keys twenty years ago, "if we saw another boat at the reef, we would be very upset," says DeeVon Quirolo, project director of Reef Relief, an action-oriented, nonprofit membership organization in Key West. "Now, at the Key West reef alone, we have 116 mooring buoys at seven sites, and they are used every day on a twice-daily schedule. Key West has the largest catamaran fleet in the world. There are six boats that carry upwards of fifty passengers to the reef twice a day. That's just in Key West. There are other operations in the middle Keys and in the upper Keys."

Coral is a delicate ecosystem, composed of millions of fragile animals called polyps with a life span of centuries. "Coral is actually an animal encased in a calcareous [containing calcium] exoskeleton," says Quirolo. "The huge coral formations are millions of colonies of these little microscopic animals encased in their exoskeletons." These colonies continue to expand year after year, with new coral building on old and passing nutrients to individuals inside. A reef is composed of the limestone remains of colonies of coral animals, each species forming a particular shape and pattern. such as leaves, fans, pipes, brains, or tendrils of plants. They grow slowly, some species only half an inch per year.

"The coral polyp is in the jellyfish family. You can crush it. When divers go hand-over-hand on the reef, they crush the corals they touch. They also introduce bacteria. We have an epidemic of black band disease. If someone were to just nick or to crush a little area of the coral and the water quality was fine, we wouldn't have that many problems. But we have millions of people just nicking and touching it." And the water quality is far from fine.

The once clear, sunlit shallows are becoming murky and green. Living corals and seagrasses are being smothered by algae that thrive on chemicals from wastewater dumped in or too near the shore waters. The chemicals, byproducts of the treatment of human sewage, are not only contaminants, but nutrients, as well. They cause tiny water algae to grow rapidly. When the algae die, the water needs increased oxygen to decompose them. The oxygen supply in the water shrinks to a level where many aquatic animals, including coral, die. Poor water quality also contributes to coral diseases such as black band, yellow band, and white band, killing corals that have endured other kinds of natural disasters for centuries.

The murkiness of the water can also be traced to soil runoff from intensive agriculture in the Everglades and as far away as the Mississippi River delta on the Gulf of Mexico. When the Mississippi flooded so disastrously in 1993, "it actually modified the pH level of our reef, according to EPA scientists," says Quirolo. The coral is smothered by soil and damaged by fertilizers and herbicides that contribute to algae blooms. Dams and irrigation systems in the Everglades have cut off water flow that is essential to the healthy ecosystem. Fingers of slimy green water are drifting into the Keys.

The reefs have stopped growing; many experts believe that at the present rate of environmental degradation, the coral reef of Florida will be dead in a few years. Dr. James Porter, a zoologist at the University of Georgia, studied six coral reef locations in Florida between 1984 and 1991. He concluded, "Five of six reefs monitored declined in percent cover of living coral over the sampling period. All reefs declined in species number during this time." Throughout the study, there was no new growth by any of the coral species that build structures.

The reef-building corals are generally found at less than three hundred feet, where light can penetrate. Sunlight is essential to the life of these species. If pollution turns the water murky, the corals cannot thrive. Although some corals can live at great depths, having been found at nearly nineteen thousand feet in the sea, that species doesn't build reefs. Most of the corals in the world live in warm, shallow, sunlit waters—the kind people like, too.

Most common in the warmer areas of the Pacific and Indian Oceans, the Caribbean, and the Gulf of Mexico, corals live in all the oceans of the world, including the Arctic and Antarctic. One of the oldest types of ecosystems on Earth, coral reefs are also the largest living structures on Earth. Australia's Great Barrier Reef, at 1,240 miles in length, may be the largest structure ever built and certainly is the only living structure visible with the naked eye from outer space. Reefs are richly diverse in species, as rich as the tropical rainforests. Hundreds of species, including hard and soft corals, sponges, jellyfish, anemones, worms, snails, crabs, lobsters, rays, sea turtles, and other sea creatures, depend on the coral reefs.

In the Florida Keys, the complete ecosystem of the coral reef actually includes not only the reefs themselves, but also neighboring seagrass beds and mangrove forests on the coastline. "People do not fully comprehend how delicate the interdependent nature of this ecosystem is," says Quirolo. "The coral reefs depend on seagrass meadows and mangroves. It's a cycle between the three habitats. If we don't value seagrasses and mangroves as well, we will not keep our reefs alive."

Seagrasses, flowering marine plants, provide food and habitat for pink shrimp, lobster, snapper, and other sea life. They filter sediments out of the water, release oxygen into the water, and stabilize the bottom with their roots. The seagrass meadows provide a filtering system to prevent sedimentation in the sandy areas from smothering the corals. They are also habitat for many sea creatures.

The third part of the ecosystem, the mangroves, thrive in coastal areas where fresh water meets salt water. Usually between twenty-five and forty feet tall, the trees form dense, bushy stands with tangles of stiltlike roots exposed at low tide. The submerged roots provide a nursery and breeding ground for marine life that later moves onto the reef. The branches provide nesting areas for birds. Mangroves also stabilize the shoreline and filter land-based pollutants. As coastal development replaces the mangroves with canals and shorelines and resorts, the breeding ground and nursery of fish and birds that later migrate to the reef is destroyed. This breeding and nursery facility serves the reef, which in turn supplies seventy percent of the local commercial fishing industry catch and is home to one-third of Florida's threatened and endangered species.

The reef ecosystem of rock and coastal forest serves as a natural barrier against the frequent tropical storms and hurricanes that batter the many islands of the Florida Keys. Even the gentle waves and soft breezes for which the region is celebrated can erode the beaches. In other areas, where the reef has been destroyed, by mining for limestone, for example, the suddenly unprotected coastline is devastated. Allowed to remain in place, natural breakwaters of reef ecosystems have kept themselves and the coastlines maintained and repaired for eons—no maintenance crews or sea walls required.

To protect the Florida coral reef, the Florida Keys National Marine Sanctuary was created in 1990; it extends 2,800 square nautical miles on both sides of the Keys. A comprehensive management plan and a water-quality protection program are being devised for the new sanctuary. "It's the first sanctuary to have a water-quality protection program," says Quirolo. "I have been working for two years now supporting the creation of that program and trying to provide a conservation voice in the formation of that plan. It's been pitifully underfunded. We got $280,000 last year, even though Congress appropriated $4 million for it. If we had the funding, it would provide money for corrective measures for upgrading systems for sewage and storm-water runoff, as well as for researching and monitoring water quality, a management plan, patrols at the reef—all the things that go along with protecting the environment."

Lacking such help, Reef Relief is working with the National Oceanic and Atmospheric Administration to start up a cooperative program called Reef Ranger. "We're going to go out and patrol the reefs ourselves. Try to intercept lost boaters. Try to give out educational materials. Let people know they shouldn't be standing on the reef. Try to prevent accidental boat groundings—which we have a lot of." In the Key Largo section of the National Marine Sanctuary, at the northern end of the Keys, an underwater police force slips into scuba gear to patrol sensitive parts of the reef. The vast majority of violators are doing damage unintentionally, though, so most of the police work is educational. Officers go from boat to boat handing out brochures and explaining the importance of conservation.

Reef Relief doesn't have a national office or funding from big foundations or endowments, "so it's been a struggle financially," says Quirolo. Though she works nearly full-time at Reef Relief, she earns her living publishing an annual visitors' guide to Key West that includes plenty of reef-conservation information, as well as the expected bar and restaurant listings. Her experience in publishing helps with the educational efforts of Reef Relief. "We are now printing and distributing over eight hundred thousand of our Coral Reef Ecosystem brochures every year. They're available in every resort, dive shop, information booth, and hotel in the Florida Keys. We send thousands of them out all over the world." She also publishes a quarterly newsletter for Reef Relief members.

A Miami native, Quirolo was educated at George Washington University and the University of Miami Law School, but stopped short of receiving her law degree. She came to the Keys and married a charter sailboat skipper named Craig Quirolo. "As the years progressed, we became more and more alarmed at the anchor damage we saw on the reefs. We got together some other skippers and formed a conservation group to put in mooring buoys, and that's how Reef Relief got started. People assume that Reef Relief was founded by divers, but in fact neither of us were divers. We were sailors." Craig, who serves as director of marine projects for Reef Relief, now dives in order to spend time researching the reefs, but DeeVon sticks to snorkeling.

"Once we put the buoys in, more boats could be accommodated, and they brought more people out there. So it kind of worked against us in a sense because it made it even easier to get out to the reef. Some very hard-line conservationists accuse us of catering to the dive industry. But ours was just a sincere attempt to eliminate the anchor damage. So then we had to get involved in an educational effort. We created our first educational brochure. We got funding from local contributors. We became incorporated [in 1987] and got our not-for-profit status."

Although individual divers generally respond readily to information about protecting and preserving the coral reefs, Quirolo says the diving industry has been very difficult to educate. "Most dive operators sell scuba courses and dive equipment, and it's a business. When this reef dies, they'll just move on to someplace else. The fleets here are run by people from all over the world. The dive representatives on the sanctuary council have been in denial about their role. We have tried and fought and lost—to get limits to carrying capacity, limited entry, restrictions on gear. We fought bitterly with them, and they are fighting tooth and nail. They do not want to be regulated. They do have their environmental projects and their marketing and all that, but when it comes down to the nuts and bolts of actually doing something about what they're doing wrong, they don't want to budge."

Recreational enjoyment of the reefs, a priceless wilderness experience, is intangible, but the reefs also hold some characteristics that nearly everyone agrees are valuable. For example, the reefs may have a beneficial interaction with the greenhouse effect, the warming of the atmosphere due to increased gas emissions, particularly carbon dioxide from burning fossil fuels. Carbon dioxide dissolves into sea water and becomes carbonic acid, which is absorbed in huge amounts by coral reef animals, along with many other shell creatures, when they build up their limestone cases. The reefs may indirectly reduce the amount of greenhouse gases in our atmosphere.

Another reason for preserving coral reefs has to do with the abundance of food and medicine contained in their rich biodiversity. Properly managed reef fisheries offer a bountiful harvest of food, from seaweed to lobster. The reefs also contain medicines, some known, some as yet undiscovered. We know that seafans contain prostaglandins to treat cardiovascular disease, asthma, and gastric ulcers. A reef species off the coast of Hana,

Maui, produces a chemical called palytoxin that has been synthesized and used as an anticancer drug.

Coral skeletons are so similar in structure to human bone that they can be used in transplant surgery, necessary when a bone has been deteriorated by cancer, for example. Since 1982, more than five thousand people have had coral implants, which are not rejected the way bone grafts donated by other people are.

Harvesting food and medicine from the coral reefs may be a strong argument for conservation, or may be yet another source of potential destruction. Large quantities of seafans are already being exported from India to Europe and the United States for pharmaceutical research. If the reefs are to be harvested, we will have to find a sustainable method of collection, just as we do with any other ecosystem on land or in the sea.

The reefs are not only productive, "they're just incredibly beautiful," says Quirolo. "When you go out to the coral reefs you see purple seafans swaying in the water. You see plumes. You see corals. A lot of tropical fish, moray eels, lobsters, crabs, worms. You can see a dolphin occasionally. Or sea turtles. It's just an incredibly rich and varied experience. It's utterly unlike anything on land."

FOR MORE INFORMATION:

Reef Relief, P.O. Box 430, Key West, FL 33041.

Telephone: 305-294-3100.

Reef Relief is a nonprofit, volunteer membership organization. For a tax-deductible contribution of $20, you can become a Seafan and receive a one-year membership, a quarterly newsletter, and other membership privileges.

WHAT YOU CAN DO:

At this writing, a management plan and water-quality protection plan for the Florida Keys National Marine Sanctuary is being formulated by government organizations, nongovernmental organizations, and the public. Reef Relief is attending meetings and arguing for conservation. You can support their efforts by becoming a member. Or voice your concerns about the future of this national sanctuary by writing to the Florida Keys National Marine Sanctuary Planning Office, 9499 Overseas Highway, Marathon, FL 33050.

If you dive, snorkel, swim, fish, or go boating at coral reefs, remember that coral is a living creature. What you do (or don't do) can make a difference to the survival of the coral reefs. If you take training to learn to dive, insist that your instructor teach and practice environmentally sensitive techniques. In the water, remember that even the lightest touch with hands or equipment can damage sensitive coral polyps. Snorkelers should wear float-coats to allow gear adjustment without standing on the coral. Divers should use only the weight needed and practice proper buoyancy control. Even lifeless areas on the ocean bottom may support new growth if left undisturbed. Avoid wearing gloves and touching or collecting marine life—most tropical fish that are captured die within a year, including those commercially gathered for aquariums. It is illegal to harvest coral in Florida, and buying it in shops only depletes reefs elsewhere.

When boating, remember that dumping trash at sea is illegal. Avoid grounding accidentally or otherwise on the reef, or dragging your propeller in shallow seagrass beds. Use reef mooring buoys or anchor in sandy areas away from coral and seagrasses. Use a holding tank for sewage, and dump it at pump-out facilities. Use biodegradable bilge cleaner, and never discharge bilgewater or sewage over a reef.

COLORED COTTON

Sally Fox, Wickenburg, Arizona

A HOBBY GONE WILD—that's what started Natural Cotton Colours, Inc. The company's founder, Sally Fox, a teenager in the San Francisco Bay Area during its high-hippie era of the late 1960s and early 1970s, loved to spin "anything I could get my hands on— silk, linen, mohair, cotton—even my dog's hair." Beyond her spinning wheel, she was also very interested in ecology, the environment, and organic gardening. The two interests intertwined.

Now she heads her own company, growing her own breeds of organic cotton that come off the plant in earthy shades of brown and green, as well as pure white. The colored cotton is woven into cloth for T-shirts, hats, clothing, or bedsheets made by such major manufacturers as Levi Strauss, Esprit, and Fieldcrest. Early into the enterprise, Fox trademarked her cotton, calling it Fox Fibre™. The fiber does not need to be dyed or bleached, a major environmental advantage, because both of those processes require large amounts of toxic chemicals, including heavy metals used to fix dye to cotton fabric. Nor does Fox Fibre fade with washing; instead it rather magically does the reverse and actually intensifies in color with washings. "It's great to pull your clothes out of the dryer, and they actually look better than before," says Fox.

Fox, who earned her bachelor's degree in biology and also has a graduate degree in entomology, began her professional life working in integrated pest management—the science of controlling pests without the use of pesticides. While working for a cotton breeder who happened to have a few brown-cotton plants, Fox fell in love with the color. Most wild cottons actually bear colored fibers rather than white. In fact, colored cotton has been cultivated for thousands of years. In India, brown cotton fiber was used to produce the original khaki cloth.

Cotton seeds are generally planted in the early spring. The plants bloom after four months, but the flowers fall away quickly, in a few days. It takes about another month for the seed pods, or bolls, to become fully grown. The pods burst open to reveal a cloud of soft, downy lint surrounding the seeds. The cotton gin, invented in 1793 and the basis for gins today, separates the cotton lint, or fibers, from the seeds.

At the time Fox first saw brown cotton, "most cotton breeders ignored colored cotton," she says, "because the fibers were too short for the spinning machines." Her old hobby came to the fore, and she started trying to spin the cotton. Her love of gardening led her, in 1982, to plant a patch of colored cotton on a small plot of land in the vast, agricultural paradise called the San Joaquin Valley of California. The first cotton she harvested, she ginned (separated the lint from the seeds) by hand. From her first crops, she selected seeds from specimens with the longest fibers and strongest colors to replant. A few generations of cotton crops later, the selection process gave her the first machine-spinnable colored cottons. She also cross-bred colored plants with long-fiber white cotton and produced some interesting new varieties that way. After a few years, she made another big discovery: "I found green lint on two of the brown plants. I realized green was a recessive gene of brown-colored cotton." Now she had two colors and bushels of enthusiasm. "I worked on it all the time. I took vacations in my cotton fields. All my friends thought I was crazy. They don't now."

By 1988, she had expanded her small plot into many acres, and her cotton fibers were long enough that Fox could sit down at the spinning wheel and spin them successfully. Fabric mills began knocking on the door. The Japanese, extremely sophisticated textile consumers who value colored cotton, were the first to buy. Two years later, Levi Strauss and Co. approached her and became her biggest customer. Now, she has left her job in biological pest control to work as entrepreneur, farmer, plant breeder, mill expert, textile researcher, fashion consultant, and globe-trotting marketer.

Fox Fibre is a hit with clothing manufacturers because it meets two top consumer demands: it is a natural fabric, and it is less expensive than traditionally grown cotton. "Nearly half the cost of preparing cloth goes to the labor and energy of dyeing and the disposal cost of toxic dye wastes," says Fox. She has been unable to really take advantage of this economic advantage, though. Her operation is too small; she is currently growing only about five thousand acres of colored cotton. The overall industry grows fifteen to seventeen million acres of cotton in the United States. "The cost of my business is so high, it puts a big weight on the price. But as soon as the acres go up, the price goes really low. But even at these current prices, you still can save money."

Fox Fibre meets another demand: fashion designers like the warm, rich colors, from red-brown to celadon green, a color spectrum that Fox says will always be limited. "Cotton, like wood, is mainly cellulose and has the same molecules that give wood its colors," says Fox. "You'll see the same color range in cotton that you see in wood—everything from the greens of the young wood to yellow, to red, to ebony." It can be mixed with white fibers to create a wide variety of brown and green shades without using bleach or dyes.

Fashion designers also like the way fabric made from Fox Fibre feels and folds—slightly different from ordinary cotton, softer and more elastic. It invites touching. It weaves into beautiful denim and chambray, and makes sheets and flannel with a softer feel than those made from ordinary cotton. It can be woven or knit, in plaids or stripes. A velvet and a corduroy are in development.

Most cotton farmers don't spend much time in the artsy offices of fashion designers. The farmers usually work with the mills, which design fabric that they show to designers. But in this case, Fox the cotton farmer designs many yarns and fabrics herself, in order to demonstrate the capability of the fiber. "Then I go to Levi's and say, 'Look what you guys could do with this.' Then the designers say to the mill, 'I want this cotton.' And the mill doesn't know how to do it, so they call me. It's all kind of backward right now."

One of the most amazing qualities of brown Fox Fibre is that it is naturally fire retardant. This unique characteristic can significantly minimize the chemicals necessary in manufacturing products such as children's sleepwear, automobile seat covers, and products made for chemically sensitive people. It passes fire standards for airplane upholstery without any further treatment, says Fox. "It passes for children's sleepwear with seventy-five percent less fire-retardant chemical than is usually applied." Theoretically the fabric could be dyed or printed any color and still be less chemically treated than most other fire-retardant apparel.

Fox's colored cotton breeds have not been readily accepted by the rest of the cotton industry. She began fighting Big Cotton because of a 1925 law that prohibits California farmers from growing any cotton other than the pure-white, high-quality Acala variety. The intent of the law was to establish the region as a premium cotton-growing district that buyers could rely on.

Although in 1990 the law was amended to allow the cultivation of Pima cotton, another high-quality white cotton, growing colored cotton in commercial quantities was still forbidden.

Fox struggled to get permission to raise cotton commercially, but she made little headway with the San Joaquin Valley Cotton Board. Finally, in 1993, Fox moved her headquarters to Arizona and the difference in business climate quickly became clear. "The first week we moved in, we received a letter from the Department of Agriculture in Arizona welcoming us as a business and telling us how proud they were to have us here. They even included the name of the person who gives out low-interest loans for new businesses moving into the agricultural areas in Arizona."

Cotton is the world's largest nonfood crop and accounts for half of all textiles. The business revenue generated by the cotton crop in the United States comes to fifty billion dollars—more than any other field crop. Each year five billion pounds of cotton is spun and woven in mills in the United States. Most of the fabric is made into clothes, but it also finds its way into hats, handbags, diapers, curtains, tents, lace, fishnets, coffee filters, bookbindings, bandages, and sutures. Procter & Gamble created Ivory soap from cottonseed oil—and Crisco shortening, too. The oil is also used in margarine, salad dressing, and cooking oil. Cottonseed meal is used to make fish bait, cattle feed, and organic fertilizer.

Cottonseed is potentially a valuable food source for people, but, unfortunately, cottonseed meal contains a toxic chemical that can't be refined out. The chemical can possibly be bred out, although so far such a species has proven hard to grow. Otherwise one of the most nutritious vegetable seeds in the world, cottonseed contains all nine essential amino acids, the building blocks of protein. Fans of cottonseed meal like to point out that if you replaced a fourth of the wheat found in a hamburger bun with cottonseed meal, it would have more protein than the hamburger.

With cotton farming so widespread, developing methods to grow it organically is important. Currently, cotton growing uses more pesticides, weed-killers, and defoliants than any other agricultural crop. In California alone, some six thousand tons of pesticides and defoliants are used on cotton each year. Across the country, cotton farms spend five hundred million dollars a year on pesticides, encouraging an industry that is detrimental to the environment. Pesticides have to be shipped to markets, sometimes great distances away. The truck or train that has an accident along the way can wreck lives with spilled cargo. And once pesticides have been spread across many acres, the empty containers, bags, boxes, bottles, and tanks must be handled as hazardous waste—although they seldom are. Most farmers and households toss empty pesticide containers in the garbage, and they end up in landfills where they may leak chemicals into our water supplies. The pesticide manufacturing plants also create toxic waste that is difficult to dispose of safely.

Fox insists her farmers grow all her cotton strictly organically, although she's had a hard time convincing the contract farmers that organic cotton farming really works. "The first farmers I contracted with, back in 1989 and 1990," says Fox, "when I said organic to them, they would practically throw me out of the room. They considered it a really dirty word. It was very difficult to even get them to consider it as an option." Gradually, they've found that Fox Fibre plants are hardier than other cotton, and easier to grow organically.

Organic cotton farming was standard fifty years ago. "Insects were controlled by maintaining beneficial populations of predator insects, by applying natural soaps and oils, and by allowing some insect damage to occur. Weeds were controlled by cultural methods and hoeing. It took a great deal of skill and a degree of luck to produce a healthy crop. When the first pesticides appeared, they seemed to be a gift from heaven. Use mushroomed."

At about the same time, a cotton-picking machine became widely used. Now that cotton farmers had chemicals to control weeds and a harvesting machine, they no longer had to have large labor forces. "Two guys—a farmer and his son, or two brothers— could take care of one thousand acres." Earning an average of fifty to one hundred dollars per acre, two people made a decent living, and small cotton farms, heavily dependent on chemicals, became common.

Now, though, the toll of chemical dependency is becoming apparent. Pests have built up resistance to pesticides, requiring farmers to spray more often or use stronger chemicals. The pesticides also kill off many beneficial insects that prey on harmful bugs. As a result, the fields have no natural protection when new pests invade.

"A cotton farmer today can control pests fairly easily without pesticides," says Fox, the former biological pest-controller. "There are all kinds of natural ways to control pests, and good professionals who can help the farmer succeed." For example, when the pink bollworm infected California cotton fields and rapidly began chewing through livelihoods, farmers tried chemical after chemical, but the bollworm resisted them all. Finally, the farmers realized that the bollworms hid at the base of the cotton stalk during the winter. By plowing the plants back into the ground in the fall and waiting ninety days before starting to grow cotton plants again, the bollworm was brought under control.

Because cotton is not consumed as a food, many chemicals are commonly used on it that are unsafe for human consumption. But even though no one eats the cotton bolls, farmworkers and people living in nearby areas are adversely affected. Studies have shown that the chemicals routinely drift miles with the wind. When defoliants are sprayed on the cotton fields in the San Joaquin Valley, residents experience nausea, diarrhea, and throat irritation. "When I first moved to the San Joaquin Valley," says Fox, "I was sick all the time. They spray fields right next to your house. They spray right next to the schools. I want to live on a farm and not automatically give myself cancer."

Trademarking her colored cotton has been a large business endeavor and expense, says Fox. "Every time I get some money ahead, I register in another country. It's very expensive if you use lawyers—figure about $2,000 for every phrase or name you want to register—but for me, it's necessary. It would be comparatively easy for someone to dye cotton and try to sell it as natural-colored."

To help protect herself further, Fox refuses to sell her seeds. She fears that her colored cotton might be grown carelessly somewhere and the seeds escape to neighboring white cotton. "If I sell little bags of seeds here and there, and little pockets of it are grown, then the regular industry could become worried. And I believe that for colored cotton to become an industry, it has to be accepted by the cotton industry as a whole." So she doesn't want to give them any cause for complaint. Secondly, she's concerned that if she starts to sell her seed, it could be easily pirated by a large company. Even though she could sue, she doesn't have the money to succeed in a lawsuit against a large company.

Her seeds could be stolen, of course, and there is already evidence that it has happened. "Shortly after I made my proposal to the cotton board, all of a sudden some good old boys had colored cotton. Suddenly. . . just like mine. . . . They never had bred it before, and then they just had acres of it. When I asked them where they got the seeds, one of them said, 'I can take them from you if I want to.' I ended up suing them and at least they stopped growing my cotton for a while. But it cost me a lot of money.

"I'm trying to hold onto this industry as mine. I want it to be something I can control. I decide who I let grow the cotton, and of course, I pick people who are willing to try organic farming, who want to be good stewards of the earth." Fox doesn't sell to just anyone, either. She insists that products made from Fox Fibre be manufactured in countries where the labor force is well paid. She prefers to sell to American companies because they are more likely to require United States environmental standards, even in their offshore operations. "American firms bear the greatest environmental protection costs and take the greatest responsibility."

FOR MORE INFORMATION:
Natural Cotton Colours, Inc, P.O. Box 66, Wickenburg, AZ 85358. Telephone: 602-684-7199.

WHAT YOU CAN DO:
The American people hold tremendous power as consumers. We can help change the future just by shopping for, asking for, and buying products produced in environmentally sensitive ways. More and more industries are beginning to proudly display environmental certification labels. You can find everything from organically grown foods and fabrics to dolphin-safe tuna, to books (such as this one) printed with soy-based, nontoxic inks. Read the labels, know what you're buying, and support environmentally responsible businesses.

SEWAGE SANCTUARY

Bob Gearheart, Arcata, California

*T*HE TOWN OF ARCATA lies three hundred miles
north of San Francisco, on the rugged, foggy
Pacific coast. Like many college towns, Arcata's
cultural, intellectual, and political life is shaped by the local college,
in this case, Humboldt State University, which has the nation's
largest undergraduate department of environmental engineering.
The citizens and the public officials of Arcata, population fifteen
thousand, care about their environment and are willing to experiment with solutions to environmental problems, so much so that
the town's mayor affectionately describes the place as "eco-groovy."
A recent front-page article on Arcata in the *Wall Street Journal* said,
"everything from the sewage system to the local garment factory is
designed with the environment in mind."

The sewage system is unique. It looks like a vast park, but it
purifies the town's daily batch of approximately two and a half
million gallons of raw sewage. When you can turn sewage into
scenery, you're on to something; it draws admirers from as far away
as Japan.

On the surface, the Arcata sewage treatment system is 150 acres
of wetlands park at the edge of Humboldt Bay, which empties into
the Pacific Ocean. A sanctuary for people as well as wildlife, the park
includes meandering streams, cattail-covered estuaries, and
freshwater ponds, edged with groomed paths and park benches.
Until 1986, the area was an eyesore: a garbage dump, an abandoned

logging mill, some degraded pastureland, and a mound of dirt locally called Mount Trashmore.

Near the new park, the town's raw sewage gushes into an enormous, open-air concrete tub. About half of the flow has been flushed from toilets; most of the rest runs out of kitchen sinks, showers, bathtubs, and washing machines. Ninety-nine percent of the flow is pure water. The other one percent is noxious and potentially dangerous solids of fecal matter and food wastes. In the tub, something called primary treatment occurs: the solids are removed, and the remaining wastewater is drained off and channeled into forty-five acres of oxidation ponds that were built in the 1950s and, with some recent upgrading, still operate excellently.

The oxidation ponds are the site of secondary treatment. Here, more solids settle to the bottom, while bacteria and algae break down the organic compounds. "These are large solar reactors, essentially," says Bob Gearheart, one of the designers of the system and a professor of environmental resource engineering at Humboldt State. "The sunlight generates a lot of algae on the surface. The algae produce oxygen as a byproduct of photosynthesis. The oxygen is then used by bacteria to break down the wastewater. There's no energy that goes into this, other than the energy from the sun. This solar reactor is taking the place of a lot of mechanical aerators that are normally used to stabilize wastewater. It takes about two to three months, depending on the time of year, for the wastewater to move through these ponds."

Before passage of the federal Clean Water Act of 1972, water from secondary treatment, such as oxidation ponds, was considered clean enough to dump directly into streams, lakes, and oceans. When more stringent water quality regulations came along later in the 1970s, a state agency proposed an expensive, high-tech facility for Arcata. This little town was a particular concern because seventy percent of the oyster harvest in California comes from Humboldt Bay, into which Arcata's treated sewage flows. Although Arcata supported stricter water quality standards, the town didn't support costly, high-tech solutions. Out of the college came plans to hook up a natural water-purification system to the existing oxidation ponds.

"This was an innovative, alternative type of treatment system," says Gearheart. "So we had to take on the Regional Water Quality Control Board and the State Water Resources Control Board. It was a battle that lasted about five or six years. We had to go to a lot of public hearings. The city officials put a lot of effort into it."

Arcata's system works like this. Water from the oxidation ponds flows into three two-acre treatment marshes, where cattails, bulrushes, and other vegetation feed on organic matter in the water. The plants filter out certain pollutants and break down harmful nitrogen and phosphorus compounds. After a few days in the marshes, the water is pumped back into a treatment plant for chlorination, which kills any remaining pathogens.

The chlorination is required by the state, although some experts think it is unnecessary. "Time is the best killer for pathogens [disease-causing micro-organisms]," says Gearheart. "After thirty days in the oxidation ponds, natural disinfection occurs." The chlorine probably doesn't do any harm; the only objection, really, is that chlorination is an unnecessary step involving a strong chemical. The water is then sent for a final "polishing" in the enhancement marshes, thirty-one acres of parkland called the Arcata Marsh and Wildlife Sanctuary.

Though human-made, the irregularly shaped marshes with small islands look very natural, and essentially they are. First, workers sealed off the old county dump with clay. Next, they dug out a five-foot-deep lake and used the fill to make three islands in the center. They built a few dikes, scraped off some topsoil, poured some fresh water into the area, and let nature go to work.

As passive systems, the enhancement marshes require little maintenance and no energy other than sunlight. In the enhancement marshes, the sunlight grows water plants rather than algae, which is not wanted at this stage; the plants help prevent algae from overproducing. "When the plants cover the water completely," says Gearheart, "then you know sunlight's not getting into the water to produce algae. Instead, the microbes—the bacteria and fungi—are predominating." The underwater microbes attach themselves to the plant stems and eat organic nutrients in the water, which is actually already clean by official standards. This is the final "polish." The microbes are nature doing its work, twenty-four hours a day, seven days a week.

The first such project in the United States, the Arcata wetlands system has been closely monitored since it went into operation. "There was a lot of skepticism," says Gearheart. But going into its tenth year, it has performed as well as hoped, and better. By the time Arcata's wastewater flows into Humboldt Bay, it is clearer and cleaner than the water already in the bay. The system is estimated to have saved at least two million dollars in capital costs, compared with the high-tech system that had been proposed. It also continues to save the town significant amounts that would have been needed for operating and maintenance costs. And it provides a large new park that opens the public's views of Humboldt Bay. Though the system is land-intensive, the multiple overlays of features and functions provide the town with superlative public land use.

"One nice aspect of a wetland treatment system," says Gearheart, "is its elegant simplicity. It's self-regulated. There's nothing you have to do to turn anything on or anything off." This aspect is particularly appealing because most sophisticated mechanical sewage treatment plants are extremely complicated to operate.

Another advantage of wetlands/wastewater systems is that their life cycles are longer than those of mechanical plants, where "you have to replace everything in twenty years," Gearheart says. "But with oxidation ponds and wetlands, the life cycle is more like forty to sixty years. At that point you might have to go in and rebuild the dikes and do some serious replanting."

Gearheart also particularly likes that the wetlands have provided better access to Humboldt Bay. "Twenty-five years ago, people could only get to the edge of the bay by going down to the dump. Now you can go right to the edge. You can look across the bay. You can look up in the mountains. It's a nice spot. The thing that's really interesting and pleasing to me is, I see parts of the bay going back to the way they were one hundred years ago. It's a form of restoration ecology that is occurring here." At last count, the town had added ninety-six acres of wetlands and twenty-two acres of adjacent uplands to its open space.

The Arcata wetlands attract waterbirds, including ducks, coots, egrets, herons, hawks, avocets, and pelicans. In all, the number of bird species has doubled and is now more than two hundred, making the marshes a prime spot for birders. Other wildlife lives in the thickets nearby. People stroll and jog along the paths. "It's even more than a park," says Gearheart. "It's really a human sanctuary. People love to go to the marsh. We've got about 120,000 to 150,000 people a year who go down there. We've had memorial services and weddings. The hospice uses it. People picnic there." Most of these people have no idea that the park exists because of sewage.

Gearheart finds this amusing, after all the public battles ten years ago. "I think if you tried to take that marsh away now, you'd have a rebellion."

Wetlands are defined as any place that has been wet long enough to develop specially adapted plants and animals. They include swamps, bogs, mudflats, floodplains, riverine bottomlands, and marshes, from Minnesota's moist prairie depressions to the vast Florida grasslands and California coastal sloughs. They occur in every region and climate in the United States—or rather, they used to occur. By the mid-1970s, according to the Environmental Protection Agency, the United States had lost more than one hundred million acres of wetlands—an area roughly the size of California—or about half the wetlands in the forty-eight contiguous states. Alaska still contains 170 million acres of wetlands, most of it in pristine condition.

Between the 1970s and mid-1980s, according to the U.S. Fish and Wildlife Service, some 2.6 million acres of wetlands—an area twice the size of Delaware—vanished. Some three hundred thousand acres continue to disappear each year. Most of the loss is to cropland, but a good deal of it turns into urban development. Historically, communities saw wetlands as wastelands, mucky messes filled with disease and pests. The general philosophy was drain them, fill them, pave them, and put them to use.

Now we know that wetlands have much to offer us in their natural state. Wetlands rival the tropical rainforest in productivity. A report by the Conservation Foundation says, "Wetlands sustain nearly one-third of the nation's endangered and threatened species. They provide breeding and wintering ground for millions of waterfowl and shorebirds every year. Coastal wetlands provide nursery and spawning grounds for sixty to ninety percent of United States commercial fish catches."

Wetlands also exercise a unique ability to filter and purify water, whether by design, as in Arcata, or not. The plant and animal life in the saturated soil of wetlands break down pollutants. The chemical and biological processes in a wetland have the potential to remove virtually every kind of contaminant.

Any town with both oxidation ponds and ample public, low-lying wetlands—or former wetlands—can use a wetlands park to process sewage. Gearheart estimates that such a facility requires an average of twenty to thirty acres per ten thousand people. Of course, the community has to plan for growth, as well. In Arcata, the wetlands sewage system has the capacity to grow from the current two-and-a-half million gallons to roughly six million gallons of sewage per day.

A similar project is already in operation in Orlando, Florida, with a population of one million, and in some fifty other municipalities nationwide. Others are being planned in towns from New York to Arizona to Hawaii. Historically, wetlands occurred in every state of the country, and they can be restored.

The wetlands system in Arcata has provided benefits besides water treatment. The methane gas generated by the primary treatment is burned for power. The sludge left in the primary treatment is mixed with green waste, composted, and put on the city forests. Gearheart wants to make use of the fertility of the marshes as well. He's proved that cattails can produce fuel, and he can envision a day when the fuel is used in cars for the local police force, or in a cogeneration plant to produce electricity. It may sound eco-groovy now, but in another twenty years it will be different. . . . Come back and see.

FOR MORE INFORMATION:

Public Works Department, Arcata, 736 F Street, Arcata, CA 95501.

WHAT YOU CAN DO:

You can create less wastewater by practicing water conservation. Fix leaky faucets—they can waste gallons of pure, fresh drinking water—and install low-flow toilets and showerheads. Gearheart recommends toilets that use 1.6 gallons of water per flush and showerheads that flow at 2 gallons per minute.

Everyday cleaning, polishing, and painting products contain contaminants that can eventually find their way into the water supply, through our wastewater. Household items such as drain openers, oven cleaners, insect sprays, chlorine bleaches, spot removers, glues, and wood preservatives all add up to a considerable source of pollution.

Learn the names of ingredients that are harmful, such as phosphorous in dishwasher detergent, and read labels before you buy. Try using substitutes, such as detergent with very low phosphate levels. In the case of unavoidable chemicals and paints, buy only as much as you need. Dispose of them in your local hazardous waste disposal facility, rather than pouring them down the drain or dumping them in the trash.

SHAMAN PHARMACEUTICALS

Lisa Conte, South San Francisco, California

*T*HE RICHEST ECOSYSTEMS in the world, the tropical rainforests, occur where there are the poorest populations—the Third World. The tropical rainforests contain half the world's gene pool, and yet, in Brazil, where one-third of the world's rainforest grows, biological riches don't always mean economic success. Some twenty-five million children and young adults—the country's future—live below the poverty level. Poor and developing countries exploit their forests in desperate (and unsuccessful) attempts to earn the money they need. Every second of every day a tropical forest the size of a football field is destroyed; that destruction affects the entire planet, because the rainforests act as the lungs of the earth. In addition, the destruction of rainforests shrinks the habitat of animal species and contributes to their extinction.

Developed countries have also plundered the rainforests. But some have become increasingly aware of the importance of preserving these rich areas. The most potent argument for saving the tropical rainforests—the argument that appeals best to human self-interest—is their pharmaceutical richness. Prescription drugs that treat eighty percent of humanity and sell for tens of billions of dollars each year contain compounds from rainforest plants. One-fourth of the prescription drugs in the United States contain at least one ingredient taken from plants. New miracle drugs may be living in the forests, not yet discovered, because experts estimate that as few as one percent of the more than 250,000 flowering plant species have been examined for medicinal uses. Examined by scientists, that is. Some of their medicinal uses are already known to many of the forty million human beings who live in the world's forests. Traditional healers in the Amazon rainforests alone use some six thousand different plants to treat and cure a wide range of disease and discomfort, according to *Newsweek* magazine.

Their knowledge is as threatened as their ecosystem. As development invades their lands and traditions, indigenous knowledge disappears. "There's so much knowledge and tradition passed on by word of mouth, from generation to generation," says Lisa Conte, chief executive officer and cofounder of Shaman Pharmaceuticals, which describes itself as the first drug company to research and develop natural, botanical pharmaceuticals based on traditional usage by indigenous people. "As those cultures get wiped out or are fiercely influenced by what we call civilization, all of a sudden they want aspirin and Michael Jackson. They don't want their own traditions. And there's going to be nobody left who remembers. We're going to lose them forever."

In 1900, Brazil was known to have 270 Indian tribes. Today, ninety of them have completely disappeared. Scattered and assimilated into other populations, they have abandoned their ways or lost their land. One of the most obvious signs of the loss of traditional ways is the population explosion that results when people have lost traditional methods of birth control. In some tribes, women who used to bear an average of five or six children now often have more than ten.

Shaman (the word means medicine man in northeast Asian cultures) Pharmaceuticals works with traditional healers in tropical rainforest-dwelling communities to find and develop new plant-based painkillers and medicines for viral and fungal diseases. At the same time, the company is dedicated to supporting the biological and cultural diversity that produced the plants and knowledge of them. The unique approach of the company relies on ethnobotany, an interdisciplinary study of human relationships to plants that combines anthropology, botany, ecology, economics, and medicine. The field of ethnobotany was only recently developed by Harvard professor Richard Schultes, who confers his intellectual blessing on Shaman Pharmaceuticals. The company has assembled a staff that includes most of the small number of ethnobotanists in the world. The firm also hired Steven King, formerly chief botanist for Latin America at the Nature Conservancy, to head the Ethnobotany and Conservation division of the company.

The discovery process at Shaman Pharmaceuticals begins with the knowledge of traditional healers—shamans, midwives, elders, and witch doctors—to pre-select potential new drugs. "The biggest reason why products fall out of clinical trials is that they have adverse side effects that you don't want to face," says Conte. "We feel we're more likely to get successfully through the clinical trial process because the starting material has been used by people for thousands and thousands of years."

A Shaman team that includes an ethnobotanist, a Western physician, and a local contact who speaks the language makes an expedition into the deep forests of a country such as Ecuador, for example, where people live in houses on wooden stilts, with palm-thatched roofs.

The scientists come prepared to work. Already well-versed in regional diseases, they know something about the local remedies, as well as the kinds of plants found in the area, and they bring pictures of disease symptoms to show the healer. Pictures accurately depict specific symptoms, such as certain skin lesions that might elude translation. They meet with the town leader and the shaman. As the shaman studies the pictures, the scientists, trained in observation methods, carefully avoid steering the healer in one direction or another. Slowly, the shaman tells them what plants he or she uses for such maladies. They find out from the shaman which part of the plant—the leaf, bark, root, or twig—is used to prepare the potion or poultice, and how the preparation is concocted. Later, the scientists will collate this information with similar tips from medical experts in other tropical countries, such as West Africa, Papua New Guinea, and Peru. Before they leave the village, the team collects specimens to take back to California with them. Also before they leave, the physician visits patients the shaman hasn't been able to help. Occasionally, the physician can dispense some knowledge or potion from his or her own country that cures the patients or eases their symptoms.

The scientists take their plants back to the Shaman Pharmaceuticals lab, a modern facility that adheres strictly to regulatory procedures laid down by the Food and Drug Administration (FDA). In its first, dazzling success the lab identified the pure compound of a medicinal preparation that the company put through preclinical development and soon had in clinical trials. The product, called Provir™, is an oral medicine to treat respiratory viruses, including one that attacks young children. The discovery has significance in a broader sense because modern medicine currently has few drugs to treat viruses. Antibiotics treat bacteria and are sometimes used on viruses but with little effect. Shaman soon followed the first success with a second product, called Virend™, a topical agent for herpes. Herpes afflicts more than thirty million people in the United States, with another half million new cases diagnosed annually.

"In the first two years of operation we had two products in clinical trials, which is the fastest timeframe for any startup pharmaceutical or biotech company," explains Conte. "It was a strong validation that we had developed a company, not just around one or two products, but around a process that can continue."

The company carefully studies the availability of supplies and the environmental impact of extracting them. Both of their first potential drug products are based on "a tree in the Croton genus. It's like a weed, sort of a pest in the rainforest. These products are probably the most highly valued nontimber rainforest-harvested products out there," says Conte. "We've also looked into putting the tree on plantations as a backup, but it's not our intention to do that right now. We have estimated that if we produce a successful product, we would use less than one-tenth of one percent of what is naturally occurring in the rainforest." The world market for the two drugs Shaman has in clinical trials is estimated to be worth $500 million a year when the products hit the market in 1998.

The company, founded in 1989, has not generated one penny of revenue yet, but its prospects look so good that it has raised twenty-seven million dollars in venture-capital financing and funds from a corporate partner. The initial public offering of shares in 1993 brought in another forty-two million dollars. This is despite the fact that major investors and mutual fund managers thought at first that going down and talking to witch doctors had to be a joke.

Revenue, though, is still a long way off. "We're five and a half years old, and we've never made a cent. Yet, we just built into the cost of doing business returning something to society, reciprocity, and sustainable harvesting, the whole green aspect of our business. If we can do that, certainly other companies—that make money— can do that."

Green businesses such as Shaman Pharmaceuticals are cheered on by environmental activists, who hope that sustainable enterprises will lead governments and individuals away from destructive industry, such as logging and cattle ranching, the two most lucrative commercial enterprises in the economy of most developing forest countries. Studies in Peru and Belize show that sustainable harvesting of resins, oils, fruits, and medicinal plants—which can be collected without bulldozers or chainsaws—could bring in two or three times more money than farming on cleared land. In Tanzania, the export of honey from forest bees is several times more valuable than the timber produced in the country's forests. One environmental group has estimated that within twenty years the retail market for the nuts, resins, oils, and medicinal plants of the Amazon rainforest alone could total fifteen billion dollars a year.

Linking rainforest conservation to commerce presents special concerns, though. If conservation focuses on profitable forms of nature, perhaps those will be the only forms preserved. The value of biodiversity lies in the great variety of animals and plants and their unique relationships that create an ecosystem.

The two most immediate concerns center around how best to extract products from the ecologically complex rainforests—which we know we do not understand—and how to deal ethically with native people who may be unaccustomed to northern countries' concepts of materialism and property rights. The Brazil-nut business, for example, which generated great hope as a sustainable industry, developed and boomed seemingly overnight. Now, though, observers are beginning to see an absence of Brazil-nut seedlings and saplings, which may be a result of overharvesting, or may be a natural cycle that was never noticed before. We also do not know what part the missing nuts play in the food chain of the birds, mammals, fish, and insects that live there. Neither the forest dwellers, with extensive knowledge of their local ecosystems, nor modern biologists have experience in devising sustainable systems for Brazil nuts, or for anything else in the tropical rainforests. Nor do we have ethical systems established for compensating the local people. Exploitation occurs. Brazil-nut gatherers, for example, are paid about four cents per pound, about one percent of the retail value of the product, by the three companies owned by three cousins that control seventy-five percent of the market.

Despite these serious philosophical and practical questions about harvesting the tropical rainforests, the fact remains that poverty and environmental degradation are moving faster than even the most concerned activists. The world cannot wait for long-term ecological studies. Enlisting the local people and assuring them a fair economic incentive currently appears to be the only real hope for preserving both the biological and cultural diversity of the rainforests. By concentrating efforts on products that bring very high prices for very small quantities, such as fragrances and drugs, some of the abuses associated with extractive industry may be minimized. One study estimates that the full development of medicinal plants from tropical countries could earn their economies some nine hundred billion dollars a year. Third World researchers, however, are unable to pursue commercial plant remedies, because their poverty prevents them from competing with multibillion-dollar international corporations.

Shaman Pharmaceuticals was founded with noble goals. Lisa Conte says her company is "a conservationist's dream come true." Shaman takes a low-tech approach in a high-tech industry and is committed to passing up endangered plants, paying sizable royalties from drug revenue to local communities where they work—and providing compensation just for the privilege of being able to conduct research. Each expedition's budget includes a contribution for a project requested by the local people, such as potable water systems, regular dental visits, or training for apprentice shamans. The company is not afraid to take a political stand, either. At the 1992 Earth Summit in Rio de Janeiro, when President Bush's administration refused to sign the biodiversity treaty designed to stem the loss of animal and plant species, most biotech and pharmaceutical companies backed the stand. Shaman officials criticized it as shortsighted.

Conte started Shaman Pharmaceuticals with money borrowed from credit cards. "It's really wonderful. This is the only country in the world where you can do that." At the time, she had an undergraduate degree in biochemistry, a master's in pharmacology, an M.B.A., and was vice president of a venture capital firm specializing in the medical and biotechnology industry.

One morning while waiting for a meeting, she leafed through magazines about drug development. To find the one hit product, most drug companies use high-tech spectroscopy, powerful computers, and robots that can quickly scan tens of thousands of samples. The chemical makeup of a single plant may include five hundred or six hundred compounds, and each compound may have fifty or sixty different biological activities. Using enzymes, instead of laboratory animals, a company can test as many as 150,000 samples in a year—hundreds of times more than was possible by conventional testing. Though this sort of hit-or-miss testing will result in only one compound in ten thousand that gets to market, the random screening method allows such huge numbers to be studied that the scientists are likely to find something useful.

While Conte was waiting and reading, the receptionist brought in a new stack of magazines. Conte saw a cover article on the destruction of the rainforests. "A little light bulb went off at that moment," she says. "Why not use plants that have been used for thousands of years and are therefore more likely to be safe and active?" The next day she quit her job and began living on credit cards. She spent four months investigating the opportunity to start such a company. Soon convinced she was on to something, she raised a quarter of a million dollars to begin tracking down the small supply of ethnobotanists in universities around the country. By May 1990, the company had raised $3.8 million and was officially founded. It targeted drugs to treat viral and fungal diseases and diabetes.

Shamans are best at treating conditions with obvious symptoms; they don't know much about complex pathology such as cancer. The virus we call herpes is thought of by some local healers as a kind of skin wound. Nomenclature aside, they have found plants that are used for modern cures. A rosy periwinkle used for various ailments by healers in Madagascar has been transformed into a drug called vinblastine that successfully treats some forms of a disease we call cancer. The local healers and communities received no compensation for the drug company's (not Shaman) priceless, profitable "discovery." The plant is now commercially cultivated in Texas.

Other drug companies are rushing into the Third World to collect plants. The National Cancer Institute has mounted perhaps the most extensive search. Since 1986, it has collected 23,000 samples from 7,000 species in tropical areas. From this collection, the institute has three compounds in preclinical tests.

Developing countries have begun to catch on to the value of their resources. Brazil and Mexico recently curtailed exports of plants, and other countries are tightening regulations. Costa Rica, one of the most diverse plant environments in the world, containing five percent of all plant and animal species, is negotiating with drug firms to allow plant collecting in exchange for royalties and technology. Meanwhile, good news is filtering up from Brazil: deforestation rates have been declining steadily since their 1987 peak.

FOR MORE INFORMATION:

Shaman Pharmaceuticals, Inc., 213 East Grand Avenue, South San Francisco, CA 94080-4812.

Telephone: (415) 952-7070.

WHAT YOU CAN DO:

The current species extinction rate is between one and ten thousand times what it was before human civilization appeared. The rate is far in excess of the rate at which new species are evolving. We are quickly running through the biological capital that took millions of years to create.

Commit to helping preserve the world's tropical rainforests, and support organizations and companies that do likewise. The nonprofit Healing Forest Conservancy (3521 S Street, NW, Washington, DC 20007; Telephone: 202-333-3438) is one of many worthwhile groups that can use your support to help stop accelerated extinction rates of biological species and the loss of the cultural diversity of forest people.

HABITAT FORMING

Steve Packard, Chicago, Illinois

CHICAGO—CITY OF SAVANNAS? Yes, originally, the site of the city was tallgrass plains, so beautiful that civic leaders of the early twentieth century created a vast system of preserves. Today, Chicago and its suburbs are among the best places east of the Mississippi to find the remnants of the tallgrass savanna that once stretched across the Midwest.

The rare ecosystem called savanna is composed of grasslands punctuated by trees. Precisely defined, savanna is a native grassland with tree canopy covering at least five or ten percent, and at most fifty percent. Forest, by contrast, is a closed canopy landscape. Prairies are treeless. The tallgrass savannas, which were once called barrens, oak openings, or prairie groves, are composed of a distinctive mixture of plants and animals that makes them different from either prairies or forests.

Steve Packard, science director of the Illinois Nature Conservancy, a leading ecological preservation group, says that based on information from the established texts, "we thought savanna was a prairie floor underneath trees." As he began looking into the ecosystem, he found that "in many ways the savanna is as different from prairie and forest as these communities are from each other," writes Packard.

Since 1977, he and his group have been working to restore these lands to their natural glory. As a result of their efforts, the 67,000 acres of preserve lands in Chicago and its suburbs support a rich diversity of life, including five globally imperiled ecosystems and

181 plant and animal species that are endangered or threatened in Illinois. "There's more nature—beautiful natural prairies and savannas—in the metropolitan area than there is in the rural area. When you get out into the countryside," he says, "there's almost nothing left. It's almost all corn and soybeans from property line to property line. Entire counties don't have a single acre of prairie or a single acre of natural woodland left in them."

The vast preserves set aside earlier in the century are now maintained by the Forest Preserve District of Cook County, which includes Chicago. Packard saw that some things were obviously wrong on the preserves, such as rare plant populations being killed off by brushy weeds. "Nature was dying on them," he says, and he volunteered to do something about it. The Forest Preserve at first refused to believe the parks needed his help. "Bit by bit I convinced them that we were friendly and wouldn't drive them crazy and could do something valuable."

The landscape of degraded prairies on seven sites was overrun with thickets. The small grassy, flowery, open portions, "incomplete and unhealthy," according to Packard, ranged in size from a few acres to only a fraction of an acre. "We wanted to see what we had so often read of, something that no longer existed anywhere—the rich grassland running up to, and under and through the oaks. A prairie with trees." They cut and hauled away the brush "and planted the prairie species, just like the academic texts told us ought to be there."

To Packard's dismay the plants sprouted and died. "They hated it—couldn't stand it in there." He and his group kept replanting, wondering what had gone wrong. Meanwhile, in the clearings, another group of plants and grasses popped up—cream gentians, purple milkweed, yellow pimpernels, bottlebrush grass, and woodreed grass, to name a few. "They weren't what I was looking for, and I sort of wrote them off," says Packard.

Puzzled by the failing prairie plants, he started playing detective. "I found this wonderful old text from 1846 in a very obscure magazine of the times called *The Prairie Farmer* that gave a list of all the plants of Illinois that this particular country doctor had been able to find." The doctor listed the names of species he saw in west-central Illinois as he traveled his professional rounds on horseback, when most of the land was unplowed and ungrazed. He coded the plants according to where they grew. "One of the categories was prairie, another was timber, and another was barrens," says Packard. Under barrens were the same plants that had been popping up in the clearings.

He was on the trail of something. He dug into an 800-page book of the Chicago area's plants and identified more than one hundred species associated with the plants he had earlier dismissed. He discovered that prairie plants were not the same as savanna plants. The idea brought criticism. "The academics hated it," Packard laughs. "Well, isn't that terrible. I didn't realize that it might seem obstreperous of us to even suggest such a thing, but I said, 'Maybe the texts are wrong. Maybe there's a whole natural ecosystem that has been forgotten about and nearly lost. Maybe if we do it this way, it will work. Things will grow.'"

He and his colleagues began looking for as many of these "oddball" plants as they could find. To maintain the integrity of local gene pools, they collected only seeds from plants within twenty-five miles of the area they were restoring. They knew they might have to really scratch to locate them, because some of them were very rare species on the endangered list. They found "pitiful, patchy remnants" along railroad tracks, in old cemeteries, at the edge of a golf course, "just some place where these species had managed to hang on." They collected great numbers of seeds and kept records of what they gathered.

The next trick was to determine what the seed needed in order to germinate. Different seeds have different requirements. "Some need to be kept moist, some need to be kept dry; some need to be planted right away; some need to have boiling water poured over them; some need to be infected with certain bacteria; some of them need to be sanded or filed or ground in some way. Many seeds pop when they're ripe and that's how they disperse themselves. Other seeds float on the wind. Other seeds are eaten by animals." Packard and his crew were able to learn much of this information from other experts, but they learned a lot by accident. They used to gather the seeds in a paper bag and leave the top open so they would dry out. Usually, they kept the bags at home, sometimes in their bedrooms. That was how they discovered the seeds that pop. Staff members came home and found the whole bedroom "covered with seeds. In everything, on everything, under everything. Now we put a screen on top so they don't end up in your underwear drawer. People who do this kind of work and keep the seeds at home are getting used to hearing 'Ping! Pop! Pop! Ping! Ping!' from those seeds when they are drying out. It's like there is this little creature in your room."

They combined the seeds to make "savanna mix." Packard says, "We'd occasionally sift the splendid golden seed through our fingers, counting out the rare forbs [broad-leaf plants] and grasses. We gather many tens of thousands of dollars' worth of seed every year, and many of the species are not available at any price." They planted the rare seeds and sat back to watch what would happen. Two years later, they had a thriving ecosystem, including animals that are known to live on savannas, even a rare butterfly called the Edward's hairstreak. Just a few years earlier, an entomologist had searched the site for that particular butterfly and found no trace. After the restoration of the plant community, the Edward's hairstreak reappeared. Because the rare butterflies don't repopulate areas from great distances away, Packard says, the entomologist believes "that there was a surviving population there, just a very small one that wasn't easy to pick up. But when the habitat was restored, then the butterfly much expanded its populations." Some of the other typical savanna species that have returned include the silvery-blue butterfly (which had not been seen in the state for decades), the Eastern bluebird, and the Cooper's hawk.

The Forest Preserve liked what it saw on the restored land. "Bit by bit we managed more and more lands and began to develop a network of people, which we called the Volunteer Stewardship Network, who are now working on 207 sites." Those sites represent 37,554 acres. In 1993, some 4,847 people volunteered to work on them and contributed 73,768 hours of work in the field. "Those are the ones that they report to us. You can't get everyone to keep track of these things." The volunteers come out in their free time because "almost everyone these days cares at least to some degree about the environment. There's very little that most people can do where they can see results. When you do this, the results aren't quick—these things tend to work fairly slowly—but they're physical, and they're real, and you can see them. . . . The idea of being able to accomplish something that is sort of good and true and beautiful and rich, and see it in full living color . . . to know that living, breathing animals and living, photosynthesizing plants are all here thanks to your efforts is very, very rewarding. . . . So much of environmental stuff is focused on negatives, trying to stop some negative. And when you stop it, all that happens is that the damage stops being done, but you don't see anything that much. Here you can really see it."

In the early 1800s, European settlers liked tallgrass savannas so much they destroyed most of them. They chose savannas as sites for their homes and schools. The soil was rich for farming, and the open groves were so attractive they were sometimes called "parkland." Even the preserves that they protected were sheltered from what the ecosystem needed most: fire, originally set by lightning and later by the Indians.

Without fire, prairie groves deteriorate. "Early publications of the Forest Preserve District printed in the 1920s show gracious open groves with tablecloths spread for picnics," writes Packard. "To traverse some of the same ground today would require an armored vehicle, or dynamite." Thickets drove out everything else. "An especially sad (and common) landscape features forlorn, aristocratic old oaks in an unbroken sea of buckthorn—the understory kept so dark by the dense, alien shrubbery that not one young oak, not one spring trillium, not one native grass can be found. Except for the relic oaks, whose decades are numbered, the original community is dead," writes Packard.

One of the most prominent invaders of the savanna is the European buckthorn, introduced during the days before the science of ecology. The buckthorn grew so rampantly that Packard now calls it "almost like a cancer. . . . What happens when these invading species come in is a very few species kills off the rest. And we end up with what I would call ecological degradation. It's different from what people not long ago called natural succession, where something superior would come in and take the place of what was there before." That would be an example of nature taking its course. In this instance, though, nature can't take its course, because we have thrown up too many roadblocks. "Ecologists today recognize that certain kinds of disturbance, like fire or flood, are a part of natural ecosystems for long periods of time, and without them the ecosystems collapse."

For the sake of efficiency and to restore the natural fire-dependent ecosystem, Packard and his crews began cautiously burning off the brush. "Some people have wagged their fingers at us about burning brush piles, telling us we'll sterilize the soil. And we just bring them out and show them a place where a brush pile was burned. Many of the oak system plants grow better where there has been a fire. And there are great flushes of bright, colored flowers and grasses waving in the wind."

The crews also used herbicides, another controversial technique. "If we don't use herbicides, we get massive resprouts coming back, and it means our efforts are for naught. Just a little dab will do it. . . soak right into the wood and into the cambium. . . take the energy out of the resprouting effort."

Once the savanna is restored, Packard doesn't expect it to remain preserved like a painting in a museum. "Once we get it going again it will always be changing, but it will be changing with all of those species intact that were once there."

Until recently, the main efforts of such conservation organizations as the Nature Conservancy and the Trust for Public Land have been simply to acquire land to preserve. In time, they began to see that some lands acquired at a great cost had been invaded by alien species that threatened to destroy the preserve. More and more, environmentalists are developing a new science and art called restoration ecology. In cases such as the savannas of Chicago, we can see hope of repairing the damage to the ecosystem, and even of reversing the damage and saving not only the land, but also the particular plant and animals species that live in that habitat.

"When I started out fifteen years ago, it was almost like finding a baby on the doorstep, or seeing a wounded animal or a wounded person. It just called to me. It said, 'I need help.' And I started working on this and trying to figure out how to bring these things back, release them, make them healthy again. And there's nothing that rewards me so much as being able to come out on a beautiful day to find thousands of flowers in bloom, scores of species that would be gone now. They would be gone from the site. Some of them would be gone from the state. In the long run, they would be gone from the earth. But they're not. They're here by the hundreds, waving in the wind. The hummingbirds are going back and forth. There are butterflies. Birds are singing. There goes a smooth, green snake. They say thank you, and it's a good feeling. That's an important part of why I do this. It feels right."

FOR MORE INFORMATION:

Steve Packard, Nature Conservancy, 79 West Monroe, Chicago, IL 60603.

WHAT YOU CAN DO:

Savannas are just one ecosystem that needs restoring. Wetlands, forests, grasslands, and rivers can all be precious habitats. You can find out what the natural conditions were where you live and try to restore them. The Society for Ecological Restoration (1207 Seminole, Madison, WI 53711; Telephone: 608-262-9547), organized in 1987, is committed to the development of ecological restoration as a science and an art, as a conservation strategy, and as a way of defining and celebrating a mutually beneficial relationship between human beings and the rest of nature. Though headquartered in the United States, the society is international in scope.

GURU OF THE OLD GROWTH

Jerry Franklin, Seattle, Washington

*T*HE PACIFIC NORTHWEST FOREST stretches nearly two thousand miles, from the Alaska Panhandle to the Golden Gate. It curves with the coast and roams inland as far as the strings of mountains called the Coast Range and the Cascade Range. In this forest stand the largest and oldest trees in the world. Spruce, cedar, redwood, hemlock, and Douglas fir—just name them, and you can almost smell evergreen scent on crisp mountain air. Tall? They are immense, many of them three hundred feet or more and fifty feet around. Old? They are ancient, five hundred years old, many of them, and some more than two thousand years old. From this forest, the timber industry is devoted to supplying society's demand for wood products.

In the 1950s, with the invention of the gas-powered chainsaw and improvements in transportation, logging began in earnest in the greatest forest in North America, the oldest forest on earth. Timber companies chopped it down like it was a wheat field that would come back again next spring. Today, the timber companies have cut ninety percent of the old-growth forest of the Pacific Northwest. Most of what remains belongs to the public.

Logging companies developed a practice called clearcutting, in which whole mountainsides are harvested. Typically, any woody debris left behind is burned, the ground sprayed with herbicide to kill weeds, and a single species of seedling is planted. With clearcutting, five years later there might at best be a short stubble of brush and uniformly knee-high saplings where a diverse, multistoried forest once stood. At worst there might be a bare, steep slope with thin soil baking in the sun. Some clearcuts defy repeated replantings, because the soil quickly erodes without the tree roots to hold it on the slope. Very few clearcut areas have ever returned to their former grandeur, either as timber or as scenery.

By the 1970s, environmentalists had had enough of this. They wanted, and still do, to preserve the forests and their wildlife for scientific, aesthetic, and other values. They and the timber companies began to fight over who would get what: how many acres of the old-growth forest would be devoted to preservation and how many to commodity production. It was presumed that the commodity lands would be worked intensively for high yields of wood products. The preserved lands would be totally removed from timber cutting. Each side of the sometimes hostile, sometimes litigious, sometimes violent struggle believed their objective could be met only by excluding the other use.

Dr. Jerry Franklin believed that there had to be an alternative. He dared to step into the middle. It has made him one of the most famous and controversial scientists in the Pacific Northwest. Franklin has developed techniques to integrate forest ecology into the timber industry. He is an environmentalist who considers people and their needs for wood products as part of the ecosystem, rather than as an extraneous factor. The problem is "these new concepts don't serve either group's interests," says Franklin. Every way that Franklin finds to improve logging practices weakens environmentalists' strategies. Likewise, every way that he finds to protect the forest threatens a way of life that depends on logging.

Franklin saw that realistically Congress would never be able to preserve the vast amounts of land necessary to maintain the biological diversity of the old-growth forest ecosystem. His idea was to find a way to log the old-growth forests without ruining them, to try to modify logging to allow habitat for wildlife to remain and the forest to be sustained. "There needs to be more acknowledgment in the forestry profession that what is good for wood fiber production is not always best for other forest values," writes Franklin. "Conversely, conservationists must begin moving away from preservation as the sole solution for many societal objectives. Reserved lands are needed to preserve many ecological values, but most of our forest lands, particularly highly productive and ecologically diverse sites, will be used for commodity production. Hence, management of these commodity lands is critically important to all of us."

Ecosystems are complex, more complex than we once thought. They may be more complex than we *can* think, according to one ecologist. Yet Franklin wants timber companies to consider the whole ecosystem before they begin harvesting. That frustrates the companies, accustomed to thinking of one species, one clearcut, one more mile of logging road. Traditional forestry techniques have concentrated on cutting and then regrowing trees, not perpetuating a complex forest ecosystem of more than one thousand species. The New Forestry that Franklin prescribes calls for loggers to leave a "biological legacy" of a little of everything that was there before: some logs and other coarse woody debris, some green trees of different ages and species, some standing dead snags. In all, enough trees are left behind to help control moisture and provide shelter for wildlife, both essential to the maintenance of a complex ecosystem. The stands that remain are predominantly young, but "inherit" significant numbers of old-growth trees and snags. "With the new practices," writes Franklin, "the forests after cutting resemble natural forests with a mixture of tree sizes, including some large, old green trees." The area is then left to heal for fifteen to twenty years. "Each stand of timber will take different measures to keep the balance of nature intact," says Franklin. "The timber companies may lose some immediate profits from taking the trouble to plan out the environmental ramifications of their work, but the wildlife and land have not responded well to existing forestry practices."

Franklin takes particular pride in his team of scientists for recognizing the ecological importance of woody debris and piles of brush, although he and others joke about it. They discovered that even the dead and rotting wood has a purpose: as organic mulch. The dead and downed trees return nutrients to the forest floor and create shelter for many of the forest animals, from woodpeckers to salamanders. After a thousand-year-old tree dies and falls to the forest floor, it will continue to support life for four hundred years more. The joke is on Franklin as much as anyone. "I worked around logs for fifteen or twenty years, climbing over them, swearing at them, kicking them, viewing them as a fire hazard, without ever thinking they had an ecological function."

The problem, from the loggers' point of view, is that the standing dead trees, called snags, and fallen, rotting logs are unmerchandisable and a hindrance to timber cutting and replanting. Loggers would rather get them out of the way.

Another New Forestry prescription that drives loggers crazy is the recommendation that some large, living trees be left behind. To the loggers, it's like leaving money on the table. In some cases, the prescription calls for leaving about twenty percent of the value behind. On the face of it, this part of New Forestry resembles the approach known as selective logging, a practice that was in favor before clearcutting became widely adopted. The difference between the two methods is elegant and subtle. Under selective logging certain trees were selected for removal; under New Forestry, trees are selected for retention. These may be the same trees: the best specimens of large, straight, green, living lumber. Selective logging took those trees and left the stunted, scrawny trees to reproduce. Other problems with selective logging were that it required a heavy road system (the U.S. Forest Service has become the largest road builder in the world) and frequent, disruptive cuttings.

Also essential to New Forestry is careful management of the water resources such as streams, springs, rivers, ponds, lakes, and the riparian or green areas immediately adjacent to them. One of the most common and sickening sights in a clearcut is a logging road smashing straight through a trout stream, destroying it as a water resource and habitat. Even if the loggers manage to avoid such destruction, the clearcuts and their roadways cause soil to crumble away and wash down into the creeks and streams, filling them, choking off their life, and by making them shallower, changing the water temperature on which sensitive organisms depend. To try to remedy this, in the 1960s and 1970s, the damaged lakes, streams, and creeks were scraped clean by timber companies as they left the area. This solution created a new problem: scraped waterways flowed too rapidly downstream, eroding the streambank and creating damage far below in the estuaries and bays.

Since then it has been discovered that forest streams require a certain amount of leaves and other forest litter in the water. It has been estimated that some healthy forest streams must contain more than sixty pounds of wood for every square yard of water. Today, these are called "sticky" streams. The fallen litter decomposes and provides nutrients for aquatic organisms.

In a natural forest and stream relationship, the stream requires more than "sticky" stuff. It also needs fallen logs that form dams, plunge pools, side channels, and gravel beds—important spawning habitat for salmon and other fish as well as other kinds of organisms. So in New Forestry, the loggers no longer scrape the waterways clean as they leave the area. Instead, they do the reverse. They drag the woods back into the water to accelerate replacement of the habitat. Logs and boulders are embedded along the creeks to create small dams, pools, and riffles that are much like the natural habitat.

The stream and its inhabitants need a certain number of trees left standing nearby, too. For example, underwater insects hatch and fly to the trees overhanging the stream, where they nourish themselves and die; their bodies then fall back into the water to feed baby salmon. If this cycle of life is disturbed, or if other events damage the salmon nursery or spawning grounds up in the mountains, the effects are felt by all who fish for salmon far downstream. The Northwest Indian tribes, for example, not only depend on salmon fishing for their livelihood, but their rights to the fish are part of a one-hundred-year-old United States treaty.

Today, Franklin says, "Protection of riparian habitats and fisheries is probably the most significant issue on private and state lands, as opposed to old growth or spotted owls or anything else."

The famous spotted owl has lost so much of its essential habitat in the Pacific Northwest forests that it has become an endangered species, and logging has been halted on public lands in its habitat until a solution can be found. That is why President Clinton called his Forest Conference in 1993, in which Franklin participated. After the conference President Clinton asked a team "to produce a plan to break the gridlock over federal forest management." Franklin was tapped for the team.

The plan that the team proposed and President Clinton accepted puts into reserve about seventy percent of the remaining unprotected older forests on federal land. It prohibits old-style clearcutting on federal land in the Pacific Northwest and Northern California. The style of logging that will be allowed on portions of federal land is based on New Forestry techniques.

Franklin began developing his theories of New Forestry by observing the way forests heal after natural disasters. What has often been called "the eternal forest" is actually in constant transition, Franklin noted. Fire, flood, windstorms, avalanches, volcanic eruptions, and other natural disasters frequently interrupt the growth of species' populations, yet somehow they survive. In fact, these disasters shape the Pacific forest. Douglas fir, for example, has become dominant in the Cascade region because of the frequent fires in the area. Douglas fir seeds need open ground on which to germinate. This is just what they get after a fire has burned off forest litter. Later, the seedlings like to grow in sun, which is just what they get when a fire has burned off the tree cover. As a result, after a fire, Douglas firs grow faster than their competitors and have come to define the old-growth forest. Native peoples called the Douglas fir the "real tree." Towering 250 to 300 feet, individual trees six hundred years old were not uncommon, and there are records of some twice as old.

As Franklin studied the work of natural disasters, he began to see that they skip and hop over the forest, destroying some parts and sparing others. Within this patchwork, what Franklin calls a "biological legacy" of energy, nutrients, physical structures, and even living organisms remains to give birth to the new forest. The charred trees add rich organic matter to the soil; the dead trees, in the form of downed logs and snags, offer homes to small animals and birds; and the remaining green trees help to relieve the stark climate of the burned-out site and provide more habitat for birds and animals. The concept of this biological legacy became the base for his New Forestry.

Immediately after a disaster, birds that flew to safety far away, hoofed animals that escaped by running, and burrowing creatures that survived by digging deep down into the ground, return to resume their lives. Even after the 1980 eruption of Mount St. Helens, an explosion with as much force as multiple atomic bombs, where the scorched slopes resembled a moonscape, Franklin and other scientists were surprised to see how many small creatures and plants returned almost immediately. As soon as the earth was cool, gophers and other burrowers popped to the surface, pushing soil over the ashes. Birds flew back, bringing seeds caught in their feathers. Flowers such as fireweed, so called because it is one of the first to return after a fire, peeked above the layer of soot, offering food for deer. Charred logs and other woody debris became home to small rodents. Franklin saw that the forests that appear after a natural disaster are complex and rich in tree structures and animal life. They differ greatly from the dramatically simplified ecosystems that result from intensely logged forests.

Franklin began arguing that in the framework of the forest, the disruptions of logging need be no more detrimental than natural disaster, if the logging is conducted so that it closely imitates natural disaster. New Forestry can use ecological principles to create managed forests that simultaneously provide for commodity and noncommodity values. "One reason that old-growth forest ecosystems are so valuable as wildlife habitat," writes Franklin, "is the varied structures of such forests—trees of all sizes, down logs on the forest floor, large snags, and multilayered leaf canopies that extend from crown to ground." New Forestry took as its objective "the creation of managed stands which have higher, often much higher, levels of structural diversity than under current practices."

Thanks in large part to the New Forestry principles, there is every hope of maintaining the habitat of the spotted owl while allowing harvest of high-value trees.

The spotted owl, whatever its remaining numbers and no matter how important it is to you that it be preserved or sacrificed, is now, in many ways, more a symbol than a living creature. The little bird represents a complex ecosystem that even the most astute scientists—or, perhaps especially the most astute ones—admit they don't completely understand. But the good news, Franklin says, is that his techniques allow the logging of spotted owl habitat, which was previously considered impossible. "These owls are living in stands of mixed structure," says Franklin. "We can re-create those stands." If he's right, we can even recreate those stands relatively quickly. He points to the Olympic Peninsula of Washington, where the owls live in multiaged forests created by windstorm and fire some seventy to ninety years ago. By contrast, present forestry techniques such as clearcutting require 200 or even 250 years to regenerate into a habitat that attracts the spotted owl.

Franklin's personal as well as professional life has been shaped by the forests. His middle name is Forest. The son of a pulp mill worker, Jerry Forest Franklin grew up in Camas, Washington, on the deep gorge of the Columbia River, near the Gifford Pinchot National Forest. In the early 1940s, when Franklin was a boy, this area was one of the great natural wonders of the world. The evergreen stands and old-growth forests covered the mountains and valleys as far as the eye could see: rolling forest green, winding river gray, infinite sky blue. He went camping with his family and with the Boy Scouts in the forests near his home and at Mount Rainier.

By the age of eight, a small boy in the cathedral light of the towering evergreens, Franklin felt a wonderful and awesome connection to the forest that he's never lost. When he hikes through the woods, he whistles, yodels, names the birds, talks to the animals, and points out each new little flower. He notices the slightest change in the seasons and still feels awed by them. For five decades, Franklin has hiked in the forest, studied the forest, and eventually, fought for the forest.

When Franklin was on the staff of the University of Washington, Seattle, he broached the idea of a "new forestry" at a conference. He spoke out just as the political and legal battle over the spotted owl was heating up. Environmentalists had also begun Project Lighthawk, a group of volunteer pilots who flew public officials and media representatives over the Northwest backcountry forests so they could see for themselves the rapidly increasing clearcuts. Politicians and writers began telling the public about what they had seen. Soon, the cutting had shorn so many mountainsides that even casual observers on commercial airplane flights over the Northwest could see the devastation. One clearcut in British Columbia was so large it could be seen by the naked eye from space.

Meanwhile, on the Hoh River near Forks, Washington, cutting on steep slopes resulted in disastrous mudslides in the winters of 1989 and 1990. The river tributaries and their salmon-rearing habitat were destroyed. One scientist at the University of Washington estimated that the area wouldn't fully recover for 150 years.

The public was ready to support any reasonable alternative to clearcutting. Franklin was ready to start the process of reform. But the Forest Service and the timber companies were not ready even to listen.

"Industry considered it irrelevant until it impacted their

operations," says Franklin. That impact hit in 1988, when environmentalists got to both industry and the Forest Service. They took the U.S. Fish and Wildlife Service into court and forced it to give the spotted owl protection under the Endangered Species Act. Shortly afterward, the Forest Service was humiliated in court when it was shown that they had ignored the research in formulating a plan to protect the spotted owl.

Meanwhile, timber industry jobs were decreasing because the forests were running out of big, old trees to cut and because much of the harvested timber was being sent overseas. Looking for a handy scapegoat, the loggers began blaming the environmentalists and the spotted owl.

Gradually, the people in the Forest Service became receptive to Franklin's ideas. The Forest Service began an ambitious testing program in the Shasta-Costa unit of Siskiyou National Forest in Oregon. The Washington Department of Natural Resources has also initiated a pilot test.

Meanwhile, the entire timber industry has been whipped by the storm of controversy and criticism from the public, the environmentalists, and economists examining log export. Plum Creek Timber Company, for example, was the subject of scathing criticism in the *Wall Street Journal.* An article blasted the company, whose greed for profit set it clearcutting at a rate faster than the forests can regenerate, across 1.4 million acres from Oregon to Montana. The company was called "the Darth Vader of the state of Washington" by U.S. Representative Rod Chandler. Just in the Cascades operations of Plum Creek Timber Company, since 1970, the annual harvest has averaged 205 million board feet. That swelled to 300 million in the "cut-and-run" days of the early 1980s. About sixty percent of the harvest typically goes to export. In 1990, the estimated cut was 175 million board feet.

Under attack, the timber industry began making changes. Plum Creek Timber Company has been spun off from Burlington Northern Railroad. It is now a publicly held limited partnership, the second-largest timberland owner in the Pacific Northwest. And the new Plum Creek demonstrates environmental sensitivity. The maximum size of its clearcuts has been reduced from 120 to 90 acres. They are employing New Forestry on fifteen to twenty percent of Plum Creek timberland, according to a company spokesman, who said they had changed their ways because "we don't think the public will accept clearcutting." He estimates the new practice costs the company about one million dollars in lost revenue. But the money is not really lost, just in the bank, according to the company biologist, because Plum Creek can return in fifteen or twenty years, when the replanted trees have become established, and remove the trees that it saved the first time. Plum Creek is still using the "old forestry," or simple clearcuts, on a small portion of its timberlands.

The company is cooperating with state fisheries, wildlife agencies, and the native Americans who have some claim to local waters. The new Plum Creek, one of the few timber companies that actually has a wildlife biologist on staff, has also given Jerry Franklin a chance to try his techniques over a wide variety of terrain and forest conditions. Today, he calls the company "the leader in the change in forestry."

At one Plum Creek clearcut, near the town of Cougar, Washington, just south of Mount St. Helens, the company left fifteen percent of the trees uncut in corridors where elk herds and smaller animals could live and travel. Logging the area in this pattern, which resembles a mohawk haircut, cost only one to two percent more than a straight clearcut, says Franklin. The Cougar example, says Franklin, "could well become a pivotal cutting, cited in forestry textbooks of the future. It blows me away." But Franklin acknowledges that this technique "is no substitute for an old-growth forest. What we are trying to do is provide a more diverse habitat so more organisms can survive."

The Plum Creek examples are being closely monitored. "It isn't so much we know this is the answer," says Franklin. "We're going to evolve the answer over time."

FOR MORE INFORMATION:

Dr. Jerry Franklin, University of Washington, College of Forest Resources, Anderson Hall, AR-10, Seattle, WA 98195.

WHAT YOU CAN DO:

Because the public's demand for more and more wood products pressures the timber industry to cut forests greedily, you can have an effect on the future of the forests with your buying habits. Support products made from recycled materials, such as recycled paper towels, toilet paper, and office supplies. Help to create the material for these products by recycling your used paper products, such as newspapers. And encourage the development of a system for certifying wood products as having been produced in environmentally sensitive ways. One proposal is for a "Green Wood Products Marketing" label, to let consumers know that the product is not the result of destructive logging practices.

SEEDS OF LIFE

Catherine Sneed, San Francisco, California

*H*EALING THE PLANET begins one person at a time. Until environmentalists succeed in changing prevalent attitudes, behavior won't change. Yet, until we change behavior, environmental problems will only continue to escalate. Catherine Sneed started working at ground level—literally. Sneed, a counselor at the San Francisco County Jail, took prison inmates and an abandoned farm and turned up a harvest of hope.

Now, ten years later, Sneed's effort is called the Garden Project and includes the original prison garden and a few others in San Francisco's toughest neighborhoods. The annual harvest of 50,000 pounds of organic produce includes tomatoes, garlic, radishes, carrots, potatoes, snowpeas, leeks, collard greens, mustard greens, cabbage, and lettuces like romaine, Bibb, and lamb's ear, as well as strawberries, raspberries, apples, lemons, herbs, and flowers. Food from the jail garden is given away to feed thousands of homeless people, seniors, and people with AIDS. The inmates "are here because they hurt people," Sneed says. "Now they have a chance to help. That's part of the healing." The inmates who do well at gardening in prison are eligible for salaried jobs at the city gardens when they are released. There, they sell their produce to San Francisco's fanciest restaurants. The enterprise teaches the ex-convicts about legitimate business.

The city gardens give former inmates a positive place to go when they leave jail. They need to learn to work, and there's always plenty of work in a garden. Many of them say the garden has changed their lives. New garden workers soon start to linger after the end of the day, reluctant to leave. Because of the self-esteem the project has given them and the support of others trying to stay clean and sober and out of jail, many continue to come around and work even if funds run out to pay them. Fundraising is another kind of harvest that Sneed has to nurture constantly, but she usually manages to pay workers about $80 a week after taxes.

"That's enough for me to get by," says Timothy, who has been in prison for drug-related activities for nearly four of his thirty-six years and has been in the Garden Project for six months. "But I would be here even if it weren't for the money. Because I know now that I need to stay occupied with something positive. Otherwise, I'm just going to fall back in my bad ways. It's like these plants. If you don't care for them, they're going to get stunted. Then even when you do go back and start taking care of them again, they can only start growing from where you left off. They never get over being stunted. So you've just got to start from where you're at."

Timothy has learned from Sneed that the garden is a metaphor. "I take these big, giant crack dealers and show them the roses we've cut back," says Sneed. "And they say, 'These ain't no roses. These are dead sticks.' And I say, 'You watch these dead sticks. They're just like you. Get rid of the dead stuff, and the new stuff will grow.'"

They learn about nourishing growing things and that they too are growing things. They use phrases like, "I've grown as a person"— not such an unusual phrase to hear, ordinarily, but unusual among ex-convicts.

Working in the garden, they see weeds threatening to smother and crowd out healthy life. They pull the weeds and talk about what's crowding out health in their lives: drugs, prostitution, or stealing. "I started gardening," said one member of the project who had been convicted of beating his wife, "and I began to see that even if you pull all the weeds out of the plot one day, they're right back there again the next. It's just like my life. I got to keep after it if I'm going to grow."

Sneed has always kept the garden entirely organic to provide another metaphor for the workers. "Most people who come to jail are substance dependent," says Sneed. "What I wanted to show them is how much better life is without chemicals. You take a chemical and put it in the garden and you get quick results, but what does it do to the soil? I wanted to show them that it's like the quick fix you get from heroin."

Chemical pesticides and herbicides may kill not only pests, but beneficial life forms, too, such as tiny bacteria, fungi, and molds, which are essential to convert nutrients in the soil into food for the plants. Continued use of chemicals can degrade the soil, creating acidity, alkalinity, and other problems that are invisible in the soil but result in poor plants. If the plants are food-producing, the pesticides can leave residues that may be harmful to the people eating the food. Pesticides can have a detrimental effect on the gardeners who handle, spread, sprinkle, and spray them, as well as on other plants and animals in the garden, including family pets and backyard birds. The use of pesticides can even encourage more outbreaks of garden pests.

For example, aphids eat leaves and can kill a plant. Ladybugs eat aphids. Spraying pesticides to remove aphids will very likely kill the ladybugs, too. If the aphids come back again, they will not be controlled by a natural predator in the garden, and pesticides are the only resort. The garden becomes trapped in an endless cycle of heavy chemical dependence.

But the problems begin before the chemicals enter the garden and extend after they've left. At the source, the chemical manufacture frequently creates hazardous waste that is difficult to dispose of safely. Also, the manufacture is usually an energy-intensive procedure that adds pollutants, such as carbon dioxide and byproducts, to the atmosphere. Likewise, at the other end of the line, after the chemicals have been applied to the garden, rain or irrigation washes the chemicals down to underground water supplies that we depend on for clean, safe water to drink. The chemicals are washed into our streams, rivers, and bays, where they can contaminate drinking water, poison fish, and poison people who eat the fish.

A study tracking pesticide runoff from fruit orchards along the Sacramento and San Joaquin Rivers in California found thirty percent of all samples to be toxic to aquatic organisms. Ninety percent of the toxic samples contained the pesticide diazinon. Not only is diazinon applied to commercial orchards, it is also a common ingredient in household and garden pesticides. Though no known tracking studies have followed household use of diazinon, it has been estimated that home gardeners use twenty times the rate of pesticides that farmers do.

Because of these and other problems, federal law now regulates labeling of products that contain pesticides. But you have to know what to look for and how to read the fine print on the label. It is far safer to avoid pesticides altogether. As much as sixty percent of pesticides are used to keep plants looking pretty and free of blemishes, rather than to keep them alive. It is estimated that ninety percent of the insects on a household lawn are not harmful.

Synthetic fertilizer, intended to pump up rapid growth, contains chemicals such as nitrates that can be harmful when they run off into the water supply. In the agricultural San Joaquin Valley of

California, well water was found to contain nitrates from fertilizers and manure at levels so high as to pose hazards to infants, some livestock, and some crops.

Natural fertilizer can be made by composting garden cuttings. Grass clippings, leaves, kitchen vegetable refuse can all go in the compost pile. Other organic matter, such as animal remains in the form of bone meal or blood meal also can be added. The organic matter slowly decomposes, and when spread around the plants, it releases quantities of healthy, natural nutrients. Composting is one of the gardening techniques vigorously practiced at the Garden Project.

Catherine Sneed grew up under tough circumstances. She ran away from a troubled New Jersey family of fourteen children. Eventually, though, her big heart led her to become a prison counselor in the San Francisco County Jail. She felt frustrated seeing the people she counseled being released with nowhere to go but back into the same environment that had caused their downfall. "When they got out, I could help get them some clothes, maybe a little job," says Sneed. "I could find out what the pimp had done with their babies. But I was not impacting their lives. When they went back to the world, they were squashed like bugs."

Then in 1984, Sneed became ill with a serious kidney disease. She underwent two months of chemotherapy and lay in the hospital for four months. Everyone thought she would die. While she was confined to the hospital bed, "a friend brought me Steinbeck's *The Grapes of Wrath*, and it really grabbed me. These were the people I worked with! Strong, independent people just trying to make it and not having a way out. In the book, the key to the family turning their circumstances around was land, working the land. I thought, we can do this at the jail—we can start over again!" The jail had a working farm designed within its walls when it was built in the 1930s. Over the years the farm was abandoned. Prisoners didn't know how to till the land anymore. The farm became a garbage dump.

"We have to get the farm working again," she told her boss, San Francisco Sheriff Michael Hennessey, who came to sit at her hospital bedside every day.

"Sure, Cathy, do anything you want," he said, to soothe her; he was thinking, as everyone else was, that she would soon die.

"It was not clear to me I was going to live," says Sneed. "But I told the doctors I was going home." The doctors allowed her to go home—to die, they thought.

"I could barely walk," says Sneed. "My kidneys weren't functioning properly, which made me swell up to an enormous size. But Mike Hennessey had given me his promise, and I collected on it. He gave me $300 and four inmates, and sick as I was, I waddled out and started clearing that field. We had no tools, no experience. But I remember one guy, a speed addict—huge, lots of muscles, tattoos everywhere. And there was an enormous bramble bush blocking a patch of land. I said, 'If we could just move this bush, that's where I'd like to plant.' I waddled off to try to find a tool, and when I got back, this guy had started tearing the bush down with his bare hands. All he needed to hear was, 'Do this and it will make a difference.'"

"Cathy," says Sheriff Hennessey, "has a magical ability to change people's lives. She has such an ability to encourage and inspire."

Her gardens have flourished and her disease went into spontaneous remission. But the disease is always in the back of her mind and drives some of the urgency with which she lives today. "If we fail to give these guys hope, they will die. We're not just making a pretty little garden here, we're saving lives."

FOR MORE INFORMATION:

Garden Project, 35 South Park, San Francisco, CA 94109.

WHAT YOU CAN DO:

Help bring greenery, gardens, and nature into the life of someone less fortunate than yourself. Gardens can be planted on patches of ground between housing projects, beside churches, and on abandoned lots. Gardens are powerful tools for affecting people. The calming and soothing effect of greenery has been known for hundreds of years. In the early 1800s, Friends Hospital in Philadelphia pioneered horticultural therapy. The greenhouse for its patients was the first in the nation. Today, there are more than seven hundred members of the American Horticultural Therapy Association. Kansas State University offers a four-year degree in horticultural therapy.

Many studies have found that being around plants can significantly benefit each of us, and especially the physically and mentally disabled, older persons, and inner-city youth. Working in a garden, people become aware of the larger forces in the world. Stress levels fall. Studies have shown that in public areas where there are no flowers, there will be more litter. Another study, of patients in a Pennsylvania hospital, found that those with rooms with views of greenery recuperated from surgery faster and were more polite to the nurses.

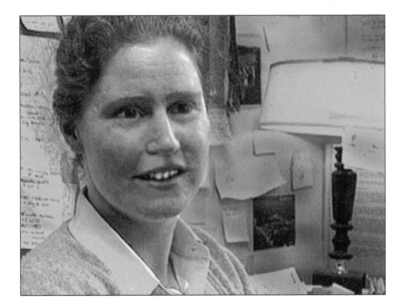

THE POPULATION CONNECTION

Nancy Wallace, Washington, D.C.

*P*EOPLE WORKING on the issue of world population frequently talk in numbers, statistics, data, and abstract concepts. But the issue itself is anything but abstract. It is essentially humanitarian. The number of people sharing the planet affects all issues about quality of life and conservation of resources.

The population issue is "all about every child on earth being planned for, wanted, and loved," says Nancy Wallace, former director of the international population program of the Sierra Club, and a lobbyist whose job involved educating politicians and bureaucrats about population issues. "The population issue has always been a pro-child, profoundly caring thing for me to work on. If you really love children, then you don't have so many that you can't give them a wonderful life. That's true for the family, the community, and the planet. Population stabilization gives people health, security, nutritious diet, and clean water." The population issue drives nearly all other system dynamics. With a sustainable, stable population, all systems work better: ecosystems, city systems, economic systems, family systems, and human welfare systems.

Population stabilization, often called zero population growth, has only three requirements. First, every couple should have no more than two children. This would lead to stable populations because each person replaces only himself or herself.

The second requirement is less well known and less quickly understood. By delaying the birth of the first child until the mother is at least twenty-five years of age, the number of generations of people on the planet at any one time is kept lower. Here is an example: if a woman has a child by the time she is sixteen years old, and her child also has a child by the time she is sixteen years old, then at the end of thirty-two years, there are three generations in the family. If the grandchild also has a child when she is sixteen years old, in forty-eight years there are four generations alive at the same time.

If, instead, the first woman waited until she was twenty-five years old to have her child, then at the end of thirty-two years there are only two generations in the family. If the child waits until she is twenty-five to have a child, in forty-eight years, there are still only two generations in the family, rather than four as there were above.

This delaying technique alone "counteracts the population buildup we've had over the last twenty years," says Wallace. "It's a fairly easy method for stabilizing population. People still have the same number of kids, hopefully no more than one per person, but simply by delaying the first birth an extraordinarily powerful statistical effect occurs. In fact, in many situations on the globe today, it is just as important numerically to delay the age of first reproduction as to reduce the number of children—it's that powerful an effect."

Because the responsibility for having fewer babies and taking care of them ultimately falls to women, the third requirement is the improvement of the status of women in general. Women who are educated and employed tend to have fewer children than women who are not, even within the same country. In countries where women have the option to be educated and employed, population growth is lower than in countries where women have fewer options.

When education and employment for women are being

instituted, Wallace feels, natural resource management and conservation should be emphasized. Improvements in resource management, including endeavors such as reforestation, drip irrigation, and no-till agriculture, help ensure that the fewer babies born will all survive to enjoy a higher standard of living. "Otherwise, the people will go back in ten or twenty years to having four or five children because half of them are dying," says Wallace. A child who lives only a few years still uses natural resources, even if only briefly.

Pause for a moment to consider some of the facts and figures:

• The world population, now more than 5.5 billion people, is growing at a rate of almost 100 million people a year. (The entire population of Mexico is fewer than 100 million people.)

• Without action now, the growth rate could double in fifty years and triple before the end of the twenty-first century.

• More than ninety percent of population growth takes place in the developing world. In industrialized nations, seven out of ten women use effective contraception. Fewer than half that many women in South Asia and Africa use birth control.

• Since 1984, world grain production per capita has fallen by one percent a year. In 1990, eighty-six nations grew less food per person than they had a decade earlier.

• In agriculturally advanced nations, there isn't much more farmers can do to increase production.

• It has been estimated that 120 million to 300 million couples in the world want family planning yet can't get access to it. To provide these couples with family planning, at an average cost of $16 per couple, might require as much as $4.8 billion—the equivalent of one week's military spending by the industrialized nations.

• Individuals don't necessarily pollute, but when there are too many people, pollution is one of the inevitable results. Overpopulation creates factors that cause global warming, species extinction, ocean overharvesting, deforestation, and desertification.

Most species die off when their breeding habits outstrip their environment's ability to support them. Human beings don't have to wait that long. If action begins now, the world's population could be stabilized at ten billion people, less than double the present size.

It is frequently assumed that religious sanctions, particularly those of the Catholic Church, are a major cause of overpopulation. Yet, Wallace notes that "about eighty percent of the Catholics in the United State actually use contraception." In developing nations, she's found that "religion is a major block only in those countries where the political leadership wants more children, usually for military purposes. But people tend to want contraception, and wise political leaders realize that a nation's economic development benefits when people have only the number of kids they can take care of. Even Iran and Bangladesh, two very religious nations, are promoting family planning. In Iran, about four years ago, the religious leaders interpreted the Koran to support contraception. The Koran says you should only have the number of children you can raise and educate in the Koran."

Another belief holds that overpopulation occurs in developing countries because those people shortsightedly consider the advantage of another pair of hands to be more important than the disadvantages of another mouth to feed—the idea being that the cure for too many children is to have more children. But Wallace finds that "where people are really suffering because they can't get food for their kids . . . or a woman is having her eighth child and her body simply can't take it, while she's also nursing at least one other child and trying to provide food every day for the first six . . . they tend to want contraception but can't get it."

Equally pervasive is the idea that the problem lies outside our own borders. Although ninety percent of the population growth occurs in developing countries, the current U.S. population of 259 million is growing by about 3 million people per year. That appears to be a small increase compared to Kenya, for example, where the current birth rate will double the population in about fifteen years. But the U.S. population uses far more than its share of world resources. By the time an American reaches age seventy-five, he or she has used five times as much energy as the world average. Even the relatively small population growth rate of one percent per year in the United States produces 25.6 billion additional pounds of carbon dioxide each year. Carbon dioxide acts like an invisible blanket to trap heat in the atmosphere and is the largest contributor to global warming.

This is just one example of the ways that individuals' use of resources—whether a small or large use—affects everyone on earth. In the United States, we need to either dramatically reduce our population, or dramatically reduce our use of resources. Or some of both.

Wallace came to the population issue after starting out wanting "to help save something wonderful in nature, like dolphins and other endangered species." While she was working as a lobbyist for the International Fund for Animal Welfare some years ago, she heard about a shorebird whose last nesting area faced eradication by American military bulldozers. The nests of the Guam rail, a flightless bird, in an area the birds had been occupying for thousands of years, lay in the brush between landing strips where President Reagan was scheduled to stop on his way to China. Security forces wanted to clear all the brush because snipers might hide there. If the bulldozers fired up and plowed through, the little bird would become extinct.

Wallace and an environmental lawyer flew into action. "We moved heaven and earth," Wallace says, "in a four-and-a-half-day campaign. We got emergency endangered species listing, which was an unheard-of thing to do. We got Secretary of Defense Casper Weinberger personally involved, as well as the Secretary of the Interior. We got a front-page story in the Sunday *New York Times*. And by Sunday afternoon, someone had talked to the head of the military base in Guam and gotten him to agree to clear the brush by hand." The nests were spared.

Wallace and her lone colleague felt exultant. They got on the phone to Guam and told the local scientist, "We've won." Wallace was surprised that he didn't share their joy. "He was just very quiet, and then he said, 'Do you think you might be able to help me save the Guam broadbill?'" He had heard the last female in existence calling in the wild and thought that unless she was netted and taken to a zoo in the next few days, she would be shot by local hunters.

Wallace flew into action again, but by ten o'clock that night, exhausted, she put down the phone and fell asleep. When she struggled awake the next morning, she realized, "We could win every skirmish and still lose this war." A hundred species are going extinct every day, faster than she could hope to save them, no matter how hard she worked. She felt mad, frustrated, and heartbroken.

"It just broke me. I said, 'something is wrong. We have to admit that we're losing.'"

Crushed, Wallace quit her job to give herself time to think. There had to be another way. "I developed a bottom-line question: if we did X, would that save the planet? And I kept testing things, working backwards: since pollution and almost all toxic chemicals come back to the oil industry, if we get rid of the oil industry, will that save the planet? And the answer was no. So I tried to go one stage back from there, and I said, if we get rid of the energy industry—coal, nuclear—will that save the planet? No.

"Finally I was sitting in a traffic jam and it came to me! We have too many people! For a couple of months I sort of tested that. If we stopped increasing the number of people, will that save the planet? And it fit. And then I started learning about it, and it was so clear. The scientists were all saying it. But the environmental movement hadn't really done population yet. Strategically, I thought, it could be the missing link, because obviously, it's not that we've tried it and it's failed. But we haven't even really talked about it yet."

Wallace turned her lobbying skills to the issue of population. Lobbying for any issue can be hard work. Lobbying involves building coalitions and keeping them "fighting for the goal, rather than fighting each other. When you're working twelve, fourteen, sixteen hours a day, it's because you care a lot. And when you spend that much time together, small differences take on a large feel. And lobbying efforts are won by small increments, so small differences can matter a lot.

"Then too, as an issue starts to succeed, a backlash grows. When you're small and ineffective, nobody cares. But it becomes harder and harder with each step closer to the goal. The last step is one hundred times harder than the first."

FOR MORE INFORMATION:

International Population Program, Sierra Club, 408 C Street N.E., Washington, DC 20002.

WHAT YOU CAN DO:

By the time an average American reaches age seventy-five, he or she has used five times as much energy as the world average. That same average person will have generated fifty-two tons of garbage. Even though the United States has nearly stabilized its birth rate, we all need to work to reduce our impact on the environment. Recently a community of 150,000 people experimented to learn if they could actually reduce pollution levels. For four days they drove cars less, at lower speeds, and they carpooled more. They also used less of other kinds of fossil-fuel-burning machines such as lawnmowers. Nitrogen oxides levels in the air dropped forty percent, and hydrocarbons came down thirty-six percent.

PRAIRIE PROPHET

Wes Jackson, Salina, Kansas

*I*N THE VERY CENTER of the heartland of America, a plant geneticist watches the grass grow and plots a revolution in agriculture. Wes Jackson struggles to find a way to feed the hungry and save the earth. He takes as his guide the prairie as it existed before the sodbusters arrived: thousands of miles of tall, deep-rooted, luxuriant grasses mixed with legumes and wildflowers—the original waving fields of grain.

When settlers plowed the prairie to plant crops, they cultivated a breadbasket, based on fields growing almost exclusively corn, wheat, or soybeans. Now, the breadbasket's got a hole in it, and our rich soil is running out. Soil erosion from American farms is measured in tons per acre—as many as sixty tons in the worst cases, and more than five tons on more than eighty percent of American farms, including most of those on the prairies. The wheat fields of Kansas and the corn fields of Iowa lose as much as two bushels of soil for every bushel of grain produced. Each year, the United States loses enough topsoil to cover the entire state of Connecticut an inch deep. Overall, one-third of America's richest topsoil, a thin layer of irreplaceable resource, has vanished since the first settlers arrived.

Soil blows away when the land is plowed for annual planting and washes away with irrigation, running off the land into lakes and rivers where it is lost forever to productive use. The land left behind is devastated by the loss, and the water is devastated by the gain. Soil runoff smothers lakes and rivers with sediment that buries small organisms and destroys delicate ecosystems. The overload of soil diminishes the amount of sunlight that reaches aquatic plants, reducing their ability to create the oxygen necessary for aquatic life.

To replenish the nutrients lost in the eroded topsoil, farmers have come to depend on chemical fertilizers. These chemicals, too, join the soil runoff and end up in waterways. Nitrogen and phosphorus from the fertilizers cause tiny water algae to grow rapidly. Soon you can see great "blooms" of algae lapping the shore. When the increased numbers of algae die, the water needs increased amounts of oxygen to decompose them. The oxygen supply in the water shrinks further, killing many aquatic animals. The plants the animals would have eaten intrude on the waterway, and it eventually dries up. Because agricultural fields today are monocultures, i.e. they contain only one crop, they lack the natural checks and balances against pests and pathogens that would be found in the ecosystem of a naturally diverse field. The chemicals in the soil can seep into our drinking water supply, making it poisonous. In Iowa and other agricultural states, nitrates have already poisoned wells in rural areas. Agricultural chemicals are the single largest polluter of our waterways, almost half of which are now classified by the EPA as damaged or threatened. Fertilizer and pesticide chemicals show up in groundwater in twenty-six states.

These chemicals not only run off into the waterways, they also drift into, and pollute, the air. Nitrogen fertilizers may be helping to thin the ozone layer. Pesticides enter the lungs and tissues of birds, animals, and people.

If all of this weren't bad enough, another environmental concern involves the machinery that runs farms and food-processing plants. These machines are heavy consumers of fossil fuels. In fact, the food the average American eats requires the equivalent of more than thirteen barrels of oil per person each year—ten calories of fossil fuel for each calorie of food. One result: air pollution due to oil consumption.

Our so-called "tending of the soil" has created environmental damage so intense that some scientists warn that human life will be seriously threatened if immediate corrective action is not taken. Wes Jackson is taking immediate action.

Believing that nature is the best teacher, Jackson studies the native prairie plants—perennials [plants that live more than two years] with constant soil-sheltering cover and soil-gripping roots. The importance of perennials in soil conservation is well known. In one experiment, the Department of Agriculture planted two fields on steep, adjacent slopes: one with annual crops and the other with perennial grass. The crop-planted land eroded at the rate of seven inches of topsoil in eleven years. On the other side of the hill, the grass-planted land stayed so intact that it was estimated the same amount of loss would take thirty-four thousand years to occur.

Perennials not only protect soil, they even help build it. Soil is a structure of dirt (mineral matter) and humus (organic matter) with a busy population of worms, insects, and other organisms that work to create and improve the soil. The prairie's falling leaves and dead plant matter create humus—natural fertilizer that also helps soil retain rainfall. In addition, humus allows air to diffuse through the soil, preventing crusting and reducing erosion. It moderates soil temperature, lessens evaporation, and discourages weeds.

The prairie's natural ecosystem balances various species of plants and animals and compensates for changes caused by weather or invading pests or pathogens, and may remain relatively unchanged by such influences. A field of corn, on the other hand, with only one dominant species, is a very simple ecosystem. It is easily destroyed by drought, insects, disease, or overuse.

The botanically rich mix of the native prairie resembles an environmentally correct factory for turning out fiber, starch, fat, and protein for livestock that can live on the leaves and other fibrous parts. It is ideal grazing land, but native prairie plants do not produce significant yields of grain (seeds) that people can eat. It is only that last factor that Jackson wants to change.

Jackson wants to develop or discover the field of his dreams: perennial grain crops based on the prairie's mix of grass, legume, and composite (a family of sunflowerlike plants), that would hold and nourish the soil, cope successfully with insects and disease, conserve water, yield at least 1,800 pounds of edible seed per acre, and run on sunlight. The grass and legume would provide starch and protein, and the composite would provide oil. The legume would release nitrogen to naturally fertilize the soil while the sunflower, experts believe, might release a natural herbicide to keep away weeds. Jackson's ideal polyculture would not just be a hodgepodge of species, though. It would have to consist of plants that have histories of growing together, in a balanced ratio of grass-ness to sunflower family-ness to legume-ness.

To develop such a perennial polyculture, Jackson founded a nonprofit organization called the Land Institute. It covers some three hundred acres in north-central Kansas, at the spot on the map where you would point if you wanted to indicate the geographic center of the contiguous United States. Since 1986, he and his small crew of researchers, scientists, and college interns have been growing approximately one hundred acres of native tallgrass (averaging 18 to 24 inches in height) prairie, sixty acres of restored prairie, twenty-five acres of woodlands, and seventy-five acres of arable bottomland. These are the revolutionary plots where Jackson and crew design and conduct experiments that they hope will lead to a sustainable system to grow food. Sustainable agriculture is defined by the Food and Agriculture Organization of the United Nations as agriculture that is "environmentally nondegrading, technically appropriate, economically viable, and socially acceptable."

Jackson says, "Since agriculture has been in existence for the last ten thousand years, we have yet to build an agriculture as sustainable as the nature we destroy. So we're looking at nature's wisdom. We're looking at the natural integrities inherent within natural ecosystems, rather than to human cleverness. So in a sense—this may sound somewhat immodest—this represents sort of an equivalent of the Copernican Revolution for agriculture." (Copernicus changed the way we think about the world and how we fit in it when he discovered that the sun, rather than the earth, was at the center of the solar system.) "So what we're saying is that modern agriculture operates under some wrong assumptions. In other words, it is almost antithetical to the way natural ecosystems have worked, and it relies primarily on human cleverness. So that's what makes it a fundamentally different paradigm. This is not just some kind of fix-it approach." At the Land Institute, says Jackson, "The philosophy informs the work, and the work informs the philosophy."

The revolution Jackson hopes to inspire would require farmers to completely change their approach—to adopt a new paradigm. The industrialized farm of today treats the whole field as one homogeneous landscape, or assembly line. The farmer plants a crop seed, fertilizes the whole field, sprays the whole thing with pesticides, and irrigates it. Unfortunately, as we are already seeing, the assembly line self-destructs. The agriculture industry is running out of ecological capital: the soil under it and the oil pumped into it. With or without Jackson's revolution, farmers are going to have to treat their fields more like works of art, each one individual and precious. Crops and farming techniques have to be adapted to the land, not the other way around. On a farm where soils differ in type, moisture, and fertility, the farmer would have to vary his plants according to where they were planted. "Beginning with an individual plant, then the field (an ecosystem), then the farm (a larger ecosystem), and then the farm community (an even larger ecosystem), it will be necessary to think about the efficient use of material and energy resources," says Jackson.

Wes Jackson was born and raised a farmer, near Topeka. He attended Kansas Wesleyan during the 1950s and later received a master's degree in botany from the University of Kansas and a Ph.D. in genetics from North Carolina State. He taught for a while at Kansas Wesleyan, began writing and publishing articles and books, and moved to California State University in Sacramento in 1971. After a few years of teaching a course on environmental and energy issues, which he called his "ain't it awful" class, he wanted to do something positive about the problems. He returned to Kansas to open an alternative school, which has grown into the Land Institute, with an annual budget of approximately half a million dollars, mostly from private foundations, although individual donations are welcome.

Jackson, being a good farmer, has the patience to watch the grasses grow—and the legumes and sunflowers—and he turns impatient when you ask how long this will take. "People want to know what we can give them right now. The answer is, not much." He says they're still at the stage the Wright Brothers were at Kitty Hawk in 1903, and he notes that the development of the airplane industry had considerable boost from the federal government's interest in defense. The equivalent governmental department, the U.S. Department of Agriculture (USDA), hasn't helped accelerate Jackson's research. "That's the thing that's so discouraging about all this. Because I think what we're talking about here is far more important than going to the moon."

The government is not ignoring the problem of soil erosion, though. One fairly new program pays farmers to keep susceptible slopes ungrazed, uncut, and planted in grass or trees. Out of four hundred million acres of till land, this new program covers less than one-tenth of them and costs billions of dollars. Soil is being saved, more than 574 million tons a year, but Jackson points out that the soil being saved is not the best; some of it is not even topsoil, because that vanished long ago, leaving the poorer subsoil.

Soil erosion today happens nearly invisibly compared to the days of the Dust Bowl, the violent dust storms that occurred during the 1930s on the southern Great Plains as a result of unwise use of the land. During the decade before, the development of the tractor, the combine, the one-way plow, and the truck led to a great plow-up on the prairies. Farmers acted on the mistaken advice of agricultural experts that repeated turning of the soil resulted in a texture that would hold and absorb rainwater better. Even when the field was left fallow, the farmers—almost like kids playing with new toys— plowed it to prepare it for future plantings, thinking that the plowing discouraged weeds and encouraged water absorption in the meantime. In the short term, the system looked good: record crops were harvested during the next few years. Everybody, including the experts, forgot about wind erosion and forgot about the drought that routinely visits these parts every twenty years or so.

The drought began in the winter of 1931–32. By spring, strong winds began to blow away dry topsoil. Spring rains were slight, followed by a few hard-hitting summer floods that battered the land and washed off more topsoil, but did little for the overall water levels. The fall was dry, and by winter many of the fields had been abandoned. The dust storms started swirling in January 1933 and blew on and off for more than four years. The sky grew dark with dust, and huge dunes formed where the dust landed. Crops were destroyed, and so were people's lives.

People began setting out for California in what was one of the largest migrations in the history of the country. In 1933, the Soil Erosion Service (soon renamed the Soil Conservation Service) was developed as a division of the USDA. These efforts did not come soon enough, though; it was like closing the barn door after the cow got loose. By 1934, desperate people of the Oklahoma Panhandle were advised by the Secretary of the Interior, Harold Ickes, to pull up stakes and move on. Later, government rehabilitation programs worked to restore the land.

To create a working model of this new paradigm, the Land Institute began the Sunshine Farm Project in 1991 to explore the extent to which a modern farm can supply its own energy, sponsor its own fertility, and produce food. The farm grows fields of grain, as well as livestock including cattle, pigs, chickens, and horses. The Sunshine Farm uses a range of renewable energy technologies, such as a diesel tractor fueled with vegetable oil from crops grown on the farm, draft horses, solar-power panels and wind turbines for generating electricity, crop rotations, and conservation tillage methods.

Jackson extends his paradigm from the farm to the community. "We need to preserve soil because that's what we're made of. If you have poor soil, you are going to have poor people." Some four million American farmers have left the land since 1944, for various reasons. Some sold out to large farm companies, some were pushed out by bad times, some were pulled away by the lure of the city. Families that once were self-sufficient, growing their own food, now have to make money to buy food.

Jackson wants to find out what kind of society could be created based on the ecological and moral values he espouses. In a small nearby town, called Matfield Green, with a population of about fifty, he has started a project to try to compare and apply the workings of a natural ecosystem community to a human community. "We begin with what we call 'setting up the books for ecological community accounting' . . . to see how it is that the money and the people disappear from these places. So your ledger includes everything from the soil that erodes to the inability of those communities to make a living there," which results in the young people leaving and the community shriveling. "Well, why can't they make a living there? In trying to get at why, we then look at what it is that the people think they need in order to be there. . . . To start with, the community needs potatoes, grains, and fresh fruit. It needs local economies, not economies from long distance. . . . It isn't a matter of a strictly local economy, but a matter of an economy that features locality."

Another important program of the Land Institute is the internships for eight to ten graduate-level students to learn with their hands as well as their heads, in the gardens and fields as well as the classroom. Interns with an interest in teaching or research are particularly encouraged, in the hope that they will gradually help change the ingrained thought patterns at the big universities. Gradually, a difference is coming about. At least seventeen state agricultural colleges now offer courses in sustainable agriculture, and several offer full-scale programs.

Agriculture based on the standard of nature is elusive, Jackson admits, but he thinks it's the best standard we've got. He has been called the "prophet of the prairie," but it's a handle he isn't much fond of. "I don't believe in prophets. Prophets arose within a royal tradition. In other words, it was a way of speaking truth to power. . . . I'm a Jeffersonian, and what I intend to do is try to make sense. And

I intend to be, if necessary, engaged in a spirited defense of a position, but I think we're all in it together and that these things have to be widely discussed and widely agreed upon. And that is a slow, slow process."

FOR MORE INFORMATION:

Land Institute, 2440 E. Water Well Road, Salina, KS 67401. Telephone: 913-823-5376.

WHAT YOU CAN DO:

You can practice sustainable agriculture in your backyard garden by making your own humus out of compost—essentially rotted plants, such as grass clippings, fallen leaves, and kitchen waste from potato skins, carrot, celery, beet tops, and lettuce leaves. Directions for making a compost pile vary, depending on your locality, and you should consult your local organic gardening store. The resulting compost makes the finest and cheapest humus around. Spread it in the garden as mulch, where it will help retain moisture and resist erosion.

Gardens and crop fields aren't the only land that erodes. Construction sites for building roads, housing, industrial sites, and airport expansions create massive disturbances that encourage soil to wash or blow away. Builders should cover excavated ground with mulch, such as straw or fiber netting. Another way to mulch is called hydroseeding; it involves power-spraying grass seed, fertilizer, and mulch on excavated slopes. The quick-growing grass then stabilizes the soil against erosion.

BARNYARD BIODIVERSITY

Don Bixby, Pittsboro, North Carolina

*Y*OU'VE HEARD OF the endangered spotted owls—but the Spotted ass? The Tamworth pig, the White Park bull? These barnyard breeds, along with the Dominique chicken, Bronze turkey, Pilgrim goose, and Cotswold sheep—many of our barnyard friends—are facing extinction. As agriculture has become increasingly industrialized, fewer and fewer species are being bred. The emphasis falls on the ones with the highest production rates.

We now have one main breed of milk cow, the familiar Holstein with black and white markings. Holsteins surpass any cow in history for milk production; a single cow can produce as much as twenty thousand pounds of milk a year. Yet the cow requires a pound of high-protein food to produce a pound of milk. A breed of cow that thrives in pastures, the Kerry cow, is being lost. There may be only a few dozen Kerrys left in the United States.

When we narrow down the number of breeds this dramatically, we risk losing the entire stock. If a single breed of cow, for example, is infected with a disease to which the breed lacks natural resistance, the disease would run through the cattle like bullets. Without a wealth of other breeds available to quickly develop, we might find ourselves running short on milk for a while.

We have one strain of poultry, the Leghorn chicken, that we depend on for egg production. The chicken, in turn, depends on stored feed. If a toxin poisons the feed stored in silos, we could be short on eggs. One of our most basic and least expensive foods would become rare and dear.

Roughly half the breeds of our farm animals are in danger of vanishing, despite the fact that biodiversity is as important on the farm as on any other landscape. "I think people understand that biological systems such as rainforests, wetlands, and prairies all need biological diversity," says Don Bixby, a veterinarian and executive director of the American Livestock Breeds Conservancy, an organization dedicated to conserving endangered breeds of livestock and poultry. Biological diversity—the variety of habitats, species, and genes that allow adaptations and evolution of biological systems to continue—is called biodiversity. "Agriculture is a biological system, too, one that we've imposed an industrial model on in this century." That model has been successful in providing large quantities of inexpensive, high-quality food—so how can we complain?

Many agricultural experts believe we've got all the breeds we need; others wonder if we even know what we've got, much less what we may need in the future. Today's farm animals seem well-suited to today's requirements, but farm science changes. Lately, we've become more aware of the far-reaching effects (and expense) of heavy use of chemicals such as pesticides, hormones, and antibiotics to produce our food. The damage these chemicals cause, such as water pollution, soil erosion, and health problems for farmworkers and consumers, makes the supposed economic benefits of industrialized agriculture look more questionable. But changing chemical-intensive practices may mean changing breeds. Many older breeds have retained traits that allow them to thrive on mediocre grassland, eat farm byproducts like whey and cider pulp, resist disease, and mate the natural way. Modern livestock, on the other hand, generally require climate-controlled housing, expensive feed, drugs, hormones, artificial insemination, and embryo transplants.

Some environmentalists question the productivity for which commercial livestock breeds were developed. The Holstein needs to eat one pound of high-protein food to produce one pound of milk. The same food the cow eats is almost suitable for people to eat directly. Why send it through the cow? The Holstein stands around being stuffed with expensive feed for a mere four-and-a-half years before its udder, heavy as a full-grown man, threatens to buckle the animal's legs. At this point, the Holstein gets turned into hamburger. But when farmers cross-breed the Holstein with a Dutch Belted, the result is a cow that produce just as much milk but that are small, hardy, highly resistant to disease, calve with ease, and graze pastures with determination. "Pasture is much, much cheaper in terms of its production costs," says Bixby, "than the cost of fertilizers, pesticides, soil preparation, planting, harvesting, processing, transporting—all of those things that are part of how we make cow food."

Another older breed, the Red Poll cow, produces milk that is high in protein, rather than high in animal fat, making it attractive for today's more healthful, low-fat diets. Just twenty-five years ago, though, high butterfat content was the characteristic we sought in milk. Fortunately, at present, if we want to change the composition of the milk we drink, we still have enough variety of breeds left to be able to do so naturally.

"Biological systems are, by definition, dynamic," says Bixby. "They never stay the same. So you have to have a range of genetic choices to meet and adapt to whatever changes come along. The science of genetics is really in its infancy right now. Biotechnology can't produce new genes; it can only move around the ones that we already have. It's important that we save as many representative genetic characteristics as we can, and those are found in a wide array of breeds. Theoretically, you couldn't have too many breeds, and at this point, we don't know how many is enough."

Cows are only one example of overspecialization. Many of the varieties of sheep that are in danger of being lost have valuable genes, too. Katahdin sheep are a rare breed that can be raised for meat with little fuss: they grow a smooth hair coat that doesn't require shearing, and they have a long breeding season. The North Ronaldsay sheep can live on the beach and eat seaweed. The Oxford Down sheep is making a comeback as one of Britain's leading breeds because it provides bigger cuts of meat.

Tamworth pigs are a hardy, high-yield, low-fat breed that requires little maintenance and grows rapidly outdoors on less food than most other breeds. Just a few years ago they were practically unknown. Now that our health consciousness makes lean meat desirable, Tamworth pigs are in demand.

Even—or particularly—the lowly chicken has been overbred. "Most of the older breeds of chickens were what we call dual purpose," says Bixby. "That is, they were raised for both eggs and meat, and generally existed on forage. They were loose around the farm, where they ate bugs, grubs, larvae, wheat seeds, scraps from the garden, and scraps from the kitchen—all sorts of things. Whatever they found became part of their diet. That's why they've been so successful all around the world.

"One of the stories that we like to tell is that Rodale Farms in Pennsylvania was looking for a nonchemical way of controlling a grub that invades apples and other tree fruits. The larvae spend the winter in the debris on the floor of the orchard, in the grass and weeds and so forth. The farm thought maybe chickens would scratch through this debris and eat up the larvae. They brought some broiler chicks. But these had been selected for very rapid weight gain on very refined and highly processed feed. These chicks would not go outside their coop. They had to be practically locked out. Once there, they certainly didn't know anything about eating grubs, bugs, and weeds, and took the first opportunity to get back inside the coop. Then the farm got Dominique chicks," one of the oldest and rarest breeds in the United States, "and they were out foraging immediately. Their whole aim in life is to find some nasty little morsel." Not only that, but they also provide brown eggs and raise a clutch of chicks each year.

Breed conservation represents a change in the relationship we have had with livestock for the past ten thousand years, during which we "improved" them. "Livestock are the result of generations of stockmen and farmers making selections of characteristics that meet needs, whether they're social or environmental or market needs at the time." A breed such as the Devon cow that provided milk, draft power, and meat was split into separate strains and then nearly disappeared as its more specialized descendants became more popular.

"Breed conservation is slightly different from wildlife conservation because breeds are a product of human selection," says Bixby, a veterinarian. "They're a complex artifact, much as architecture is or language structure or many of the other things that we cherish as part of our history." The Devon cow dates back to 1623 in the records of the Plymouth Plantation. It is the oldest recognized breed in North America. "So there's a historic reason for conservation." For some people, conserving breeds is like collecting rare stamps or vintage cars: it's history, it's fun, and it's colorful—these breeds tend to look distinctive. The Jacob sheep has four horns. Bashkir Curly horses grow their hair in corkscrews nine months of the year. The Dutch Belted ox wears a wide white belt around the middle third of its black body; it looks like a Holstein whose black and white splotches have gotten their act together.

"I think variety is fascinating," says Bixby. "The idea of living in a world where there is only one breed of cow is not acceptable to me. As more and more people live on this planet, we're going to have to be more aware of our partnership with nature and with all the plants and all the animals in nature, not just the wild ones."

The American Livestock Breeds Conservancy is a nonprofit organization that works to encourage and promote not only the rare breeds, but also the breeders—who are slightly rare themselves. There are about four thousand members, and "there is no other organization in North America that's doing this kind of work," says Bixby. "Our conservation effort is based on a network of member breeders and breed associations. There are more than 200 breed associations in North America. We're in some sort of contact and correspondence with all of them. We provide a breeders' directory so people can locate stock if they're just getting interested in a particular breed. Or if they already have them, we give them an opportunity to exchange breeding stock with other breeders."

At this point, there's very little federal support for this kind of conservation in this country, although around the world the story is somewhat different. "We are founding members of an organization called Rare Breeds International, an umbrella organization of conservation groups from around the world with about thirty-five members."

In 1990, a federal program for animal genetic conservation was established. Some efforts have begun, and a repository for animal genes is under construction. "It's called a gene bank, and it includes frozen semen, embryos, and stem cells—undifferentiated cells from various species that might be used as a source of genetic material in the future. Probably we'll never be able to regenerate a chicken, for instance, from stem cells, but we might be able to harvest particular genes that would be useful in biotechnological activity."

For some breeds, the repository provides semen for artificial insemination. "So instead of having to ship the bull around the country, all you have to do is ship the semen around. It also effectively increases the reproductive population because some of the bulls from the semen bank are already dead, and yet they can continue to produce calves and behave genetically as though they are still part of the population."

Working with the U.S. Department of Agriculture to establish a genetic conservation program, the American Livestock Breeds Conservancy has provided breed information for a computer database. "People can access this computer system and look up any particular breed to find out the characteristics, distribution, population and that sort of thing." Bixby was also one of the team of authors from the American Livestock Breeds Conservancy who wrote *Taking Stock: The North American Livestock Census* published by the McDonald & Woodward Publishing Company, Blacksburg, Virginia, in 1994.

The Conservancy has also provided information to the Food and Agriculture Organization (FAO) of the United Nations. "FAO has quite an ambitious plan for conservation in developing countries. Many of the indigenous breeds in the developing countries are being substituted or diluted by improved breeds from the northern hemisphere." An industrial animal like a Holstein makes little economic sense in the developing world. "These animals are obviously not adapted to the environment, and just because they produce mega amounts of milk in California with pelleted feed, clear running water, and machinery to clean, feed, and care for them doesn't mean they're going to do very well in Nigeria, for example, or in Arabia." Rather than put a pound of feed into a cow for a pound of milk, the Ethiopian could use the cow food directly and let the animals forage—if they would. Holsteins aren't much better at foraging than broiler chickens.

When a population of a rare breed in the United States is threatened with the slaughterhouse or is being broken up and dispersed, the American Livestock Breeds Conservancy attempts direct rescue. "We try to match them up with new homes and new jobs," says Bixby. The rare breeds aren't intended for pets, though. "These animals have been selected for a particular job, and our aim is to put them to work, not just to have them as companion animals."

Nor does he encourage putting the animals in zoos, although their being on exhibit, he admits, increases public awareness. "But it doesn't help very much directly with conservation, because they need to be reproduced, and they need to be continually selected. If there's no selection process, then they adapt to whatever environment is provided for them. So, for instance, if dairy cattle are not milked, then nobody knows what their production capabilities are, and nobody selects for those capabilities. So you get sort of a degeneration of attributes of these animals. If some breed of sheep that's accustomed to ranging very widely to select its feed from marginal range land is brought in and fed alfalfa pellets and never gets to go outside again, then you risk losing the characteristics that made that breed successful, and its offspring may really have no advantage. Eventually, you get sheep that are no longer able to forage, and that's really what has happened with many of our livestock breeds. When you choose one thing, you always reject something else.

None of these breeds qualify for protection by inclusion on the Endangered Species List because they are not truly species, but subspecies. "The species level includes horses, cattle, asses, sheep, goats, pigs—obviously those species are not endangered." The breeds are the next level of differentiation. A parallel would be, for example, tigers, a species within which there are subspecies for Sumatran tigers, Bengal tigers, Siberian tigers, Chinese tigers (probably extinct by now), each of which has genetic and physical characteristics that are quite different, one from another. The rare livestock breeds represent the genetic variations of the subspecies.

A Texas longhorn cow is as important as a bald eagle. "They're all important in that they're all here for a reason. They're all part of the web of life."

FOR MORE INFORMATION:

Don Bixby, American Livestock Breeds Conservancy, Box 477, Pittsboro, NC 27312.

Telephone: 919-542-5704.

WHAT YOU CAN DO:

It's fun, and healthy, too—for you and the environment—to select a wide range of food. Buying a variety of food encourages a variety of livestock breeds. Very simply, you can try buying brown eggs that represent a slight genetic variation, although it's not critical. In the Northeast, Highland beef is being labeled and marketed as low-fat, grass-fed beef. In the Southwest, lamb is available from a breed called the Navajo Churro, particularly well adapted to the arid climate there.

THE RECYCLERS OF CAIRO

Laila Kamel, Cairo, Egypt

*C*AIRO, THE LARGEST CITY IN AFRICA and the capital of Egypt, generates six tons of solid waste each day and recycles nearly five-sixths of it. Discarded objects and even garbage are sorted and recycled into cash, skills, and better living conditions. The program started simply enough—just because poor people found so much of use in the throwaways.

The garbage collectors are Coptic Christians who migrated to Cairo in the 1940s, fleeing drought and political vendettas in southern Egypt. Unskilled and uneducated, they became squatters and collected the city's garbage in donkey carts. They built shelters out of whatever they could find in the garbage—tin, cardboard, and plastic. They fed their pigs with the food they found, dirty and sometimes rotten, but acceptable to the pigs. They squatted on vacant land, "so the government kept evicting them every few years," says Laila Kamel, a native of Cairo who received advanced academic degrees in the United States. "Until finally, in the early 1970s, a large group of Coptic Christians moved up to a sandy, rocky area called Mokattum Hills." From there, you can see the Great Pyramids.

The squatters were finally allowed to settle in one spot because of the efforts of an advocacy group called the Association of Garbage Collectors. Though they now had a settlement, they still lived among mountains of garbage, smoke, fumes, and raw sewage. In the mid-

1980s, the Association of Garbage Collectors received a grant to start a recycling industry. "Men received machines on credit to crush plastic, grind cloth, melt aluminum, compact paper," says Kamel. "They embarked on it, started learning it, and became recyclers. And that's what you have now. An amazing neighborhood. There's no unemployment there." But there is still wretched poverty. The young girls in particular live a miserable life, "sorting rotten garbage four hours out of every day of their lives, as well as helping with the cooking and caring for younger children. And that's horrible. So something had to be done," says Kamel.

Kamel, well educated, well traveled, and well-to-do, began trying to start an informal school for the girls of the garbage village. She set up the school to be flexible, so children could attend after they finished their garbage sorting and other tasks. The curriculum she designed emphasized health and hygiene, including such basics as using a latrine. The mix of garbage with sewage is dangerous and potentially fatal to garbage pickers and, through them, to others. Garbage pickers in Peru had a cholera outbreak in 1991. The disease quickly spread to surrounding countries, including Mexico, and a handful of cases were discovered as far north as the Gulf Coast of Texas.

Kamel held classes in a nearby church. "Whenever there was a funeral, which was pretty often, we had to be evacuated," she says. "And that was the end of our school day, unless we met on the roof or went looking for an empty cave or something. So it was highly flexible and informal. You never knew what was going to happen on any given day, but we had fun with field trips, plays, skits, songs."

The garbage dumps around the houses were bad, but the pig sties were worse. "It was a very unsanitary conditon, with the manure," says Kamel. In 1987, a nonprofit group called the Association for the Protection of the Environment (APE) established a composting plant. Now the people could clean out their pig sties and bring the manure to the plant, where it was treated, refined, and sold for fertilizer. Living conditions improved somewhat.

Kamel became the Volunteer Field Director for APE in 1988, and as she looked at the community in a broader perspective, she saw that the little girls being taught at the church school were the lucky ones. "There were older girls out there going on garbage routes. They never had a chance in life. Not even a miserable little literacy school in a church." APE had already started a rag recycling program to make rugs and to provide some training and employment. Kamel took it over and redirected it to serve older girls. "Girls of puberty age, who still go out on routes or sort manually, who are illiterate and poor—those who have no option in life. I decided those were going to be my target population, and I went around recruiting them."

To get the program underway, though, she needed rags as well as workers. "I knew I couldn't afford to pay for the raw materials while the girls were in training." Kamel went around to the biggest textile mills and ready-made garment manufacturers in Egypt and asked for their rags as a donation. Most of the industry's remnants are sold for such uses as stuffing mattresses, but the manufacturers were willing to give Kamel's project some of their leftover rags.

The girls learn how to sew rags into patchwork quilts and to weave them into oven mitts, potholders, kitchen aprons, vests, and other garments. New recruits sort rags for other girls who weave at the looms. Each weaver gets an equal weight of rags each week. The girls who produce more items, or items of better quality, earn more money. Dirty items earn less, so the girls learn to wash their hands and to keep a clean loom and a clean room. In order to understand the accounting of what they've earned, the girls have become intrigued with, and learn, reading, writing, and arithmetic. Kamel uses the rag trade as an opportunity to teach basic literacy as well as life skills and enhancements such as "song and dance, drama, skits, recreation, celebration, health, and hygiene—everything. It's very flexible programming and a fun place to be. And the girls get paid for it, as well as being taken away from sorting garbage."

Their finished goods are sold at handicraft fairs and through other outlets, where the girls are sometimes taken on excursions to see their products on display. They gain pride in their work and have a chance to visit more prosperous parts of the city as respected members of the community, rather than as garbage pickers. "This is a project about self-esteem," says Kamel. "These are daughters of the Pharaohs who built the pyramids. I want them to build their own pyramids of dignity and self-esteem."

In the summer, they attend "camp" together, for fun and renewal in a place with clean air, green space, and clear skies. Each year, the project trains about 150 girls. If the girl marries before the age of eighteen, she is not allowed to continue the program. Upon graduation, the girl may receive a loom to start her own cottage industry. According to Kamel, some of the first students from twelve years ago, "when they were little, teeny kids in pigtails, today are fantastic community leaders."

Egypt has a population of fifty-six million people and a birth rate that adds another million and a half per year. Twenty percent of the labor force is unemployed. An estimated sixty percent of the people are illiterate, because although Egyptian law decrees universal education, the schools are already packed. The situation is typical of the developing world, where the United Nations Children's Fund estimates that living conditions are heading backward for one-sixth of the world's people. The United States annually sends $815 million to Egypt, and more aid comes from other countries, "yet few of the poor ever feel it," says Kamel. Many foreign aid workers don't speak the local language, a distinct disadvantage when working with people who are illiterate and living on squatters' land. Many of the squatters cannot comply with laws about registration and identity cards, and distrust strangers.

Kamel works with these people because they so desperately need help and receive so little of it. She is one of the few locals educated abroad who are willing to work in the garbage village. Kamel attended the University of California, Berkeley, where she received a master's degree, then she taught Arabic at Stanford University and at Columbia University Teacher's College, where she earned a Ph.D. in education, writing her thesis on development among Egypt's rural and urban poor. She returned to Cairo and after a short while began working with the garbage collectors. She says she has learned more from them than she did getting a Ph.D. "The poor really have so much to teach us. They are illiterate, but they learn from traditions. They have a proverb that says, 'What comes out of the land has to go back into the land.' They don't waste. From the table of the rich, through the garbage collector, to the animal, to the compost heap, back to the land to grow food, back to your table. So what I did with the rag recycling was copy the poor. I got the rags from the rich, took them to the poor, manufactured them into very high-quality expensive stuff, and sold them back to the rich. There is no waste. And that's something that I think we have to learn in developed countries. It is unforgivable the way that we waste. Not just the way we waste, the way we design production models that are so disdainful of what happens to the waste."

Debates in the United States about what to do with our waste—from junk mail to nuclear weapons—rarely seriously contemplate producing less of it. Our capacity to produce waste has outstripped our capacity to dispose of it. This oversight built into our economic system accounts for goods produced, but not for waste. In fact, the cost of cleaning up pollution is usually included as a contribution to national economic health. The GNP actually increased as a result of cleaning up the tragic oil spill in Alaska from the *Exxon Valdez*.

It you add up all the waste generated each day in the United States, it totals more than twice the weight of the average citizen. Industrial solid waste accounts for most of it: an average of 320 pounds per person per day. Gaseous waste vented into the atmosphere, such as carbon dioxide, the major contributor to the greenhouse effect, adds more than twelve pounds per person per day. Five pounds a day of our individual waste is in municipal garbage. Another five pounds per person per day comes from hazardous waste. Add some 2.3 trillion gallons of municipal effluent pumped into coastal waters annually, and then tack on incalculable, massive volumes of raw, untreated sewage known to be dumped directly into our waterways. In each of these categories, we have reached the limit of our ability to keep throwing things away. In many cases, we have already surpassed the limit.

Landfills for municipal solid waste are reaching capacity and closing down. Of the twenty thousand landfills in the United States in 1979, more than fifteen thousand are now full and closed. Hazardous waste has become so difficult to dispose of that a few years ago it was discovered that truckers were dumping liquid waste by simply opening valves and letting it trickle out onto roadways all across the country. Industrial wastewater, some 4.9 billion gallons of it per year, is dumped in coastal waters, killing once-productive fisheries. In San Francisco Bay, for example, industrial waste has fed so much mercury into some fish that people can no longer eat them.

For years, we have found it convenient to simply send our

garbage "away." Now, though, we're running out of "away" places, as New York City, which for years sent its garbage out to Staten Island, has discovered. Trash shipments now amount to forty-four million pounds per day, and the garbage dump has grown so tall it will soon require a Federal Aviation Administration (FAA) permit as a threat to aircraft. The search for disposal sites extends farther afield; some officials have actually begun negotiating with other countries, where poverty or political oppression leaves the citizens little hope of resisting. Serious efforts have been made to send municipal garbage to the Marshall Islands and to Tibet.

Some places that are "away" are beginning to refuse to become dumping grounds. In 1987, a garbage barge carrying 3,186 tons of commercial waste left Long Island and was turned away by ports in North Carolina, Louisiana, Florida, Mexico, Belize, and the Bahamas. At about the same time, a cargo ship carrying fifteen thousand tons of toxic ash from the incinerators of Philadelphia spent two years cruising the world looking for a place to get rid of its load. Where it was ultimately dumped is unknown.

The ship's toxic ash resulted from another convenient but shortsighted solution: burning the garbage. Nearly twenty billion dollars is being invested in new incinerators, even though serious health and environmental concerns regarding incineration have not been addressed. One of the most important concerns is that the ash that results is much more hazardous than the garbage. Burning concentrates some of the most toxic ingredients, such as heavy metals including lead and mercury, which do not break down, but accumulate. Because incinerated garbage is rarely treated as hazardous waste, the heavy metals are released into the environment and begin to concentrate in larger and larger amounts in animals such as the fish that we eat—or, hopefully, that we are warned to avoid eating.

The waste incinerators produce air pollution, too. The emissions include toxics such as dioxin, pollutants such as sulfur dioxide, and heavy metals such as lead and mercury. If one municipality pollutes the air, we all bear the consequences to some extent. Air pollution doesn't stay put, but drifts across state and national lines and mixes in the global atmosphere we all breathe. If incinerators in Philadelphia pollute the air, you will breathe the pollution in Peoria, as well as in Phoenix or Pocatello—or Phuket. At the North Pole, with no incinerators, factories, or freeways, air pollution in the winter and spring reaches levels equivalent to many large, industrial cities.

U.S. Vice President Al Gore has said of garbage incineration, "In effect, we have discovered yet another group of powerless people upon whom we can dump the consequences of our own waste: those who live in the future and cannot hold us accountable."

For her work, Kamel received the Goldman Environmental Prize in 1994. The prize, unofficially called the "Nobel Prize of the environment," includes a gift of $60,000. Kamel has put the prize money toward furnishing a development center in a village called Sharmoukh outside of Cairo, where she is working to promote health care and encourage the traditional skill of basketweaving among the rural women.

Recently at a conference in Tunisia, Kamel heard about many other interesting projects being started in Africa. "Recycling is becoming one of the most popular industries to attract unemployed youth, as an alternative to entering gang life, with drugs and guns." The "throwaway" people are being recycled as well as the resources. There is also a similar recycling program beginning in Pakistan. "I hope we will join hands in spreading the gospel of justice for the poor, because justice for the poor is justice for the environment. So let us not grab from each other, and above all, let's not grab from the earth."

FOR MORE INFORMATION:

Laila Kamel, 31 Montazah Street, Heliopolis, Cairo, Egypt.

Laila Kamel has written a short book titled *The Recycling Miracle of Mokattum Garbage Village*. She will send you one for US$20, shipping included.

WHAT YOU CAN DO:

Our mountains of municipal solid waste are composed mostly of newspapers and other forms of paper. Another major component is yard waste, construction waste, and organic waste, which are estimated to make up twenty percent. (One study has found that approximately fifteen percent of the solid food Americans buy gets thrown out.) Another ten percent of the garbage is plastic. The simplest and most direct way you can help reduce waste is by trying to avoid creating it in your daily routines. Attempt greater efficiency. Reuse and recycle consistently.

In order for recycling to work, not only do we need to separate our garbage and send it out to the appropriate processor, we have to create a market for recycled products. Ask for and buy products manufactured from recycled material, such as toilet paper, paper towels, toner cartridges, and plastic bags.

SOLAR OVENS

Daniel Kammen, Princeton, New Jersey

*I*N THE TROPICAL HEAT OF KENYA, they're cooking with sunshine. Stews, cakes, even fried foods come out of ovens that use no wood, coal, gas, or electricity. The cooker boils water for tea and reaches temperatures hot enough to kill germs in water or meat, yet never burns food and needs little tending. Gently simmered, the food cooks in its own juices and retains vitamins and minerals. People cook meals they're used to, in pots they already have. They use solar ovens.

The solar oven project is an example of "appropriate technology," the concept that developing countries need technologies that are appropriate to their stage of development and their lifestyle. Obviously, there is no use for energy-conserving microwave ovens if most households lack electricity, as they do in rural Kenya, where about eighty percent of the country's population lives. Though any proposal of microwave ovens for families in huts may seem laughable, don't laugh—similarly inappropriate projects in medicine, employment, and agriculture as well as conservation have been attempted and have failed.

For technology to be appropriate to rural Kenya, it must take into account the way the people live there. Rural Kenyan homes generally consist of small buildings where the same room or rooms are used for cooking, sleeping, and working. Homes rarely are connected to community sewer systems. People grow their own food as best they can, and the women spend hours a day scavenging for wood for three-stone stoves that fill the houses with noxious smoke. Each day, firewood becomes scarcer. Eighty-four percent of Africans will face wood shortages by the end of this decade, according to the United Nations Food and Agriculture Organization (FAO). Technology that can improve these people's lives must be appropriate and renewable.

Sunlight is the answer, believes Dr. Daniel Kammen of Princeton University, who directs an Earthwatch volunteer project to introduce solar technology in Kenya. Kammen's project starts by connecting with a respected local person. The local person and Kammen or volunteers demonstrate the oven and explain the theory of solar cooking to interested community members. "After they've seen our demonstration, we encourage the community to form a committee and discuss whether they want to pursue the project. If they do, we come in with materials and conduct a workshop for three or four days in which we help community members build the ovens themselves. The materials are simple and affordable—plywood, foil, glass, nails."

The solar oven is a simple insulated wooden box lined with foil and topped with glass. "But there's quite a lot of carpentry work," says Kammen. The plywood must be accurately marked and cut into nineteen pieces. Assembly calls for several pairs of hands and competence with hammer and nails. The final cutting of the door, fitting the linings, gluing the glass, and attaching the lid are all exacting tasks. "It takes a while to make each oven, and once people have done it themselves, they're invested in it. They use the ovens. And when they do, other people get interested." Kammen and his volunteers help establish local committees that continue to organize and conduct workshops after the foreigners have gone home.

Solar cooking was tried in a number of sun-splashed developing nations in the 1960s and 1970s with minimal success. Kammen says the reason earlier attempts didn't catch on had nothing to do with their appropriateness or fitness as a technology. Rather, he believes the earlier projects failed because they operated as donations, and the local community felt they were being imposed on by foreigners.

Kammen says his project "plugs in right at the bottom" level where local people can benefit. He intentionally works in the opposite way of international governments and aid agencies that have tried to assist developing countries by pouring into them millions of dollars that seldom reach people on the land. Despite massive donor aid during the 1980s, the per-capita gross national product in Kenya actually fell from $410 to $370. Over the past two decades, big technological fixes—including shipments of machinery, seed, and fertilizer—have aimed at improving agricultural efficiency, yet productivity fell by almost thirty percent.

By contrast, the beneficial effects of the solar oven project can be seen immediately and on many levels. Reducing the number of smoky cooking fires improves the health of the family. Cooking smoke is consistently cited as one of the main sources of acute respiratory infection, the leading health hazard among women and children in developing countries.

At the global level, reducing cooking fires improves the environment. More than a third of the world's population depends primarily on wood for cooking, burning nearly half of the annual world wood harvest of three billion tons. Cooking fires account for sixty to ninety percent of burning in developing nations—which contributes as much as forty percent of the global greenhouse gas emissions. Burning firewood indoors creates major health hazards in villages throughout the world.

Equally important, the wood that is burned is frequently the most scarce. The rural poor often live on marginal land, at the edges of swamps, deserts, and rainforests—the most fragile ecosystems that can least afford deforestation, another global concern and contributor to the greenhouse effect. In developing countries, as populations go up, forests come down.

Depending on cooking fires can actually contribute to the cycle of hunger. In African and Asian countries that lack firewood, dung is burned at the rate of four hundred million tons per year. That much dung could supply fertilizer to grow twenty million tons of grain, enough to feed one hundred million people a year.

Because the benefits of solar cooking are so many and so immediate, even a modest number of solar ovens can have an important impact. The Kenyan households Kammen has monitored burn an average of nearly a metric ton of fuel per person per year. "That's a lot of trees. A solar oven can cut that consumption by half."

The number of households using solar ovens remains small, because the personal approach of the project takes time. "It's no good just giving things away," Kammen says. "Better to get the idea out there first and let the people decide for themselves." He firmly believes that the introduction of appropriate technology must proceed on a small-project scale and be tailored to local needs. In three years of part-time, volunteer efforts, he has gotten 110 solar ovens in use in Kenya, and more are being implemented as fast as people can be trained with them. More than sixty communities have requested informational visits.

"This is good, simple technology. Within hours, they know everything we know about solar ovens, and they spread the word. This is really techno-transfer."

There are very good ethical and practical reasons to urge the United States and other developed nations to share their physical and technological wealth. Ethically, there is the shame of possessing so much when most people in the world possess so little. Practically, we know that if we do not share, the "have-nots" will continue to rapidly deplete their natural resources, adding to global environmental concerns. A society of "have-nots" will continue to disintegrate, and its people will continue to migrate in masses, looking for better lives. This sad scenario ultimately leads to resource wars in which we will all be the losers.

It is not that the "haves" refuse to share. In fact, the United States and many other Western nations have flooded developing countries with aid, but the dollars are frequently squandered in waste and corruption, and rarely create the desired effect.

The solar ovens that Kammen and others are introducing in Africa and other areas are based on a design from the 1950s, which maximized effectiveness of the cooker and made it appropriate not only in its energy source, but also in its materials and construction. "The solar oven," says Kammen, "is remarkably simple, just an insulated box with a glass top. It works the same way a greenhouse does. Light enters through the glass and is absorbed and reflected by the foil-covered walls." The light is converted to heat that is trapped inside by the glass. Properly made, the oven can reach 350°F in strong sunlight. "The box cooks food like a crockpot—slowly but steadily, as long as the sun shines," says Kammen.

The sunshine on Kenya amounts to about nine hundred watts per square meter, according to Kammen. The solar oven traps about four hundred watts. It can cook four kilograms (about nine pounds) of eggs, rice, fruit, vegetables, beans, or chicken in about two hours. A typical Kenyan stew, such as ugali, takes about seven hours to cook. Though solar cooking takes about twice as long as traditional methods, it frees up hours that would have been spent searching for and gathering wood. The extra time the food has to sit in the pot doesn't matter much to people who work in markets or fields close to home.

A solar oven even works under partial clouds, so long as it is constantly moved to face the sun directly. Of course, it won't work at night or on rainy or very cloudy days. At those times, high-efficiency wood-burning stoves can supplement solar cooking. Such stoves differ in the extreme from wood fires so smoky your eyes tear when you are standing in the hut.

Many previous solar ovens were designed with collecting mirrors to increase the temperature inside the box. The disadvantages of this design are that they cannot be constructed locally, they generally allow only for one vessel to be cooking at a time, and they require constant adjustment to direct the sun onto the cooking vessel—the cook has to stand in the hot sun over a blinding reflector. Generally, she quickly went back to her old smoky stove.

And it is "she" who does this work. African women perform seventy percent of subsistence agricultural labor and more than ninety percent of domestic labor. Naturally, women have become the key ingredient in the success of the solar oven project. This has come about not just because the women are looking after their own welfare, but also because Kammen and his group have begun focusing on women. They have found that women are more willing to learn new skills than men. Women have also demonstrated more patience and a better capacity to work with groups. Women act as leaders, participants, and follow-up directors who encourage integration of the new technology.

The follow-up teams make visits to pass out new "solar recipes" and cooking techniques and to help with problems that arise. If the oven can be integrated into the women's basic cooking repertoire, Kammen has found, use of it will slowly increase with time. The process of fully incorporating solar ovens into a lifestyle can take several years.

Many other countries are ideal for projects modeled on Kammen's efforts. Many developing countries are equatorial, which means they have plenty of sunshine, and could employ solar cooking at the same time that they are introduced to the concept of alternative and renewable energy. By developing social and economic programs that recognize that the resources of the planet are finite, the two seemingly contradictory goals of conservation and economic growth can coexist. Energy use is the link. Renewable energy technologies can facilitate the development of a sustainable economy in all countries—developed as well as developing.

FOR MORE INFORMATION:

Daniel Kammen, Assistant Professor of Public and International Affairs, Woodrow Wilson School, 444 Robertson Hall, Princeton, NJ 08544.

E-mail: kammen@phoenix.princeton.edu

Earthwatch, 680 Mount Auburn Street, P.O. Box 403, Watertown, MA 02272.

Telephone: 1-800-776-0188.

WHAT YOU CAN DO:

When you save energy, you save money—that means you, personally, as well as your community, nation, and planet. Alternative, renewable energy, from sun, wind, or water can now perform many fossil fuel jobs. Wind hydraulic systems, waterwheels, and solar heating and cooling usually cost less than the fossil fuel equivalent, and alternative energy is more inflation-proof, too. Even in the least favorable parts of the lower forty-eight states, far more energy from sunlight falls on a typical building than is required to heat and cool it. A supplemental solar water heater can be economical to install and can cut your heating bill, too. To experiment with solar ovens, contact Solar Box Cookers, 7036 18th Avenue, NE, Seattle, WA 98115.

ORGANIC MILK

Straus Family, Marshall, California

WE'VE GOT TOO MUCH MILK. The oversupply has dropped prices so low—with little increase in the past twenty years—that farmers have had to race to find technologies and methods that will increase production—in order to squeeze every drop they can out of the cows and bring their costs down. In the past ten years, tens of thousands of farmers who can't afford to stay in the expensive race have been forced out of business. At the same time, the federal government has spent $10.6 billion buying up surplus dairy products, mostly in the form of butter.

The Straus family owns one of the last dairy farms in Marin County, forty miles north of San Francisco along the winding Pacific Coast Highway. The farmhouse is nestled on one of the most scenic hillsides in the exquisitely beautiful county. Bill and Ellen Straus moved here when Marin was still one of the most productive dairy counties in the state. By modern dairy industry standards, the Strauses have a small farm, 660 acres, with 215 cows producing 1,500 gallons of milk per day. It is generally considered just about the minimum necessary to make a living. On a typical industrialized dairy farm—more like a factory—there may be thousands of cows, bred for voluminous milk production that depends on maximum nutrition, ample clean water, fairly cheap energy, a nonstressful climate, and high-tech equipment. The animals are basically four-legged udder bearers that mill around on concrete pads in huge buildings where they stay most of their short lives, eating super feed grown with synthetic fertilizers, herbicides, and pesticides, and shipped in from somewhere else. The expensive feed is mixed according to computer calculations adjusted to the season. Each cow wears an ear tag that another computer reads to dole out its daily ration. A single worker, with the help of high-tech equipment, can milk hundreds of cows in a few hours.

Farmers are using more and more drugs to keep the cows in shape, principally penicillin and other antibiotics that fight the udder infections the cows get from producing more and more milk, while farmers also inject the cows with bovine growth hormone to increase production. The FDA admits that the use of the hormone will increase the incidence of disease in dairy cows, which will lead to still greater use of antibiotics and other drugs. Increased use of antibiotics increases the resistance of bacteria to the drug, which makes it more difficult to keep the cows free of disease.

But it's not only the cows who get the drugs. The *Wall Street Journal* has reported that its own study of drug residues in milk, and another done by the Center for Science in the Public Interest, indicate that twenty to thirty-eight percent of retail milk samples contained animal drug residues. In August 1992, the Government Accounting Office (GAO) reported to Congress that its investigation showed that as many as eighty-two drugs that could leave residues were being used on dairy farms, that sixty-four of these were not approved for use on dairy cows, that thirty-five of these were not approved for use on any food-producing animals, and that only four of these were being tested for use in milk on a regular basis. The GAO concluded that "low levels of some animal drugs in food may produce

1) allergic reactions in persons sensitive to antibiotics;

2) the development of bacteria resistant to antibiotics;

3) the suppression of the human immune system through constant exposure to low levels of antibiotics; and

4) a slight increased risk of chronic effects such as cancer. In particular, there are reports in the medical literature of the emergence of antimicrobial resistant bacteria linked to the use of antibiotics on dairy cows, which could increase the risk of human infection."

After considering reports like this, the Strauses' eldest son, Albert, came up with the idea of creating a marketing niche for the small dairy farm by selling organic milk. Although sales of organic products topped $1.4 billion nationally in 1992, up fifteen percent from the year before, organic milk is new on the shelves. So little organic milk is sold in this country that the Straus family estimates their entrance to the market will increase the nation's organic milk supply by twenty-five percent. The rest of the organic milk in the country comes from a single cooperative creamery in Wisconsin.

"I think organic milk is the kind of direction family farms must take to survive," says Albert. "It's a healthy way to maintain land, animals, and the rural lifestyle." The Straus family had been concerned about organic farming since the 1960s when they read *Silent Spring*, the book that first alerted the general public that America was poisoning itself with chemicals. The Strauses stopped applying pesticides and herbicides, even to Ellen's rose garden, and used manure as their main fertilizer. They became increasingly concerned about sustainable agriculture and land stewardship. Albert started an Ecology Club in high school and gave his life savings at the time, a Mason jar full of coins, to support the election campaign of an environmentally aware candidate for county supervisor.

But even though the Strauses had been farming with care for decades, it took them four years to meet the state's rigorous requirements of certification for organic foods. Despite the fact that there weren't any organic dairy farms, rules existed. "It takes you thirty years to learn the state's dairy laws," says Bill Straus. "And even then you don't know them."

First, to qualify as an organic dairy in California, cows must have been eating only organically grown grass and feed for at least three years. The Straus farm has been free of chemical pesticides for two decades, so that part was easy. But cows need more than grass, and it took Albert a couple of years to find organic feed suppliers. He brings in organic corn from Iowa, buys organic alfalfa in California, grows his own silage, oats, and legumes, and set up his own mill to grind the organic grains.

Second, the cows have to be kept free of antibiotics or hormones for at least one year before their milk can be labeled "organic." To treat udder infections and reproductive problems, the farmer is allowed to use aspirin and some homeopathic remedies (a medical practice based on highly diluted medicinals). Albert could only find two veterinarians in the nation who use homeopathy, and both lived out of state. Vitamins and minerals are allowed to make up five percent of the cow's diet, and cows are allowed to be vaccinated for diseases.

The next big change is at the creamery, or processing plant, which is traditionally a shared facility for many dairy farms in the region. Organic milk cannot be processed with the same equipment as nonorganic milk because it might become contaminated. Albert established the Straus Family Creamery as their own plant to process and bottle their milk. They use recycled glass bottles, which best preserves the superior taste. The milk carries no flavor of wax or cardboard. It tastes like pure milk. Even the nonfat milk has so much flavor it hardly tastes "skimmed." The Straus Family Creamery not only sells organic whole and nonfat milk, but also butter, cream, and buttermilk.

Organic milk, produced without chemical additives, is different from raw milk, which is not pasteurized to kill bacteria. The Straus milk is pasteurized but not homogenized. A layer of cream floats to the top of the whole milk. A quart of Straus Family Creamery organic milk sells for approximately $1.25 to $1.50, somewhat higher than nonorganic milk. But the cost of production is about twenty to thirty percent more, and the entire changeover has cost the Straus family $500,000. No bank was willing to make a loan on such a risky venture, so the family raised money from friends and other family members. At this writing, they are six months into the organic operation and ahead of their financial projections.

"You have to take some chances in life," says Ellen. "The worst thing that can happen is that you fail. It's no fun to contemplate that, so you just don't act accordingly. You are going to make it. You have to have that determination and that belief that you can do it."

Bill and Ellen Straus already know something about taking chances. And succeeding. They are Jews who escaped Nazi Europe in the late 1930s. They met in America, where both of them found safety and freedom, as well as opportunities never available to Jews in Europe. "For hundreds of years, Jews were not allowed to own property in Germany," says Bill. He and Ellen both felt that owning a farm and living on the land satisfied a cultural craving hundreds of years old. "There's something about nature that's good for the soul. It makes me feel complete," Bill says.

Ellen adds, "We have had values in working on the land that we never had as children. We have been able to raise our children in these values." She can't exactly say what she means by "values," but tears well in her eyes as she looks out the farmhouse kitchen window, across the rolling hills, and remembers her four children growing up here, tending animals, crops, and gardens. She says, "It just means a great deal to us that this lifestyle be continued. That's why Bill and I fought so hard to keep family farming in Marin

County and to keep the lands protected. To have that opportunity to go on for the next generations. Because once you lose it, there's no going back."

FOR MORE INFORMATION:

Straus Family Creamery, P.O. Box 768, Marshall, CA 94940. Telephone: 415-663-5464.

WHAT YOU CAN DO:

Remember that food doesn't grow on the grocery shelf. The surest guarantee for the long-term health of the land, the food, and the people is locally produced, wholesome food. Support local farmers and organic foods.

To aid understanding of the value of farms, Marin County farmers hold an annual Family Farm Day. Farmers invite city folks to come on down to the farm by advertising in local newspapers, placing notices in environmental and conservation group newsletters, and sending announcements to schools. Local, state, and even federal officials are invited by letter or phone. Everyone gets a tour of a working farm or two, and sometimes visitors are encouraged to try their hands at milking a cow or churning butter. They also get an earful of information about what it takes to keep a farm working. A big barbecue tops off the day.

TACKLING TEXAS TOXICS

Susana Almanza, Austin, Texas

*H*UNDREDS OF PEOPLE felt ill. They all lived near the tank farm, a storage facility for millions of gallons of gasoline and chemicals. Investigation revealed toxins in the soil, water, and air of the neighborhood. EPA records, toxicologists, health officials, and even gasoline company records all confirm that a certain level of exposure to compounds such as those found can cause the illnesses described: recurrent respiratory problems, throat ailments, migraine-like headaches, rashes, nausea, and nosebleeds among adults and children. A computer analysis of more than five million Medicaid records concluded that people living in the neighborhood of the tank farm suffered. Sometimes when children played in water puddles in their yards they broke out in sores. A backyard gardener developed sores from working with the soil. Not only people were affected; grass withered and died, and the trees turned lifeless. Yet residents could not find anyone to take action.

"Everyone passed the buck," says Susana Almanza. She is director of a group called PODER (People Organized in Defense of Earth and its Resources), which means "power" in Spanish. Most neighborhood residents had Latino backgrounds, and many others were African-Americans. PODER argues that polluting industries locate in minority neighborhoods because the people there lack power. The grassroots group was formed in May 1991 by local Latino activists to organize, educate, and empower the people to fight for their right to a clean environment, with immediate attention to the tank farm. The tank farm consisted of six bulk gas terminals owned by Mobil Oil Company, Star Enterprise (Texaco), Coastal States Crude Gathering Company, Chevron USA Products Company, Citgo Petroleum Corporation, and Exxon Company USA. Some of the tanks had been in the neighborhood for more than thirty-five years. The chemicals in the tanks included benzene (a known carcinogen), toluene (which causes damage to bone marrow, liver, and kidneys, as well as birth defects), and zylene (which can damage the brain).

In December 1991, Mobil sought a permit to expand its facilities at the tank farm, stating that it would continue to emit gasoline, diesel, benzene, oxides of nitrogen, and carbon monoxide. Citizens who investigated local records discovered that because the Mobil facility had existed since the early 1970s, before emission control regulations clamped down, the emissions of benzene and other gases were already at a level well above what would be allowed by current regulations. At the time, the Mobil facility emitted forty-eight tons per year; a new permit would allow only twenty-eight tons. The emissions were known to drift over a one-mile radius, covering a residential area and at least six schools. Some residents lived within a few feet of the facility, separated only by a chain-link fence. One school was within three thousand feet.

PODER immediately began to investigate further and to ask for public hearings. At the same time, the activists from PODER joined with a coalition of African-American neighborhood associations, called EAST (the East Austin Strategy Team) and in January 1992, organized a community meeting with the residents in the immediate vicinity of the tank farm. By February, the two activist groups organized a "toxic tour" of the area for local elected officials. Representatives from the city, county, and state governments, as well as neighborhood association representatives and school leaders participated. "When they got off the bus and they could smell the odors out here from the gasoline, and they could see the conditions that the people were living in—it was a real eye opener," says Almanza. The hour-long tour included brief visits with several residents who reported numerous health problems and a close-up look at standing water and land near the tank farm.

Within days of the tour, testing of the soil and water near the fuel site began. One water sampling showed concentrations of a cancer-causing substance as high as 720 times the federally acceptable level. "The picture just kept getting bigger and bigger, and scarier and scarier," says Almanza. Next, the people were hit in the pocketbook—as a result of the contamination, the county appraiser depreciated the value of more than six hundred homes surrounding the tank farm by as much as fifty percent.

"We started organizing," says Almanza. "We established block captains and phone trees and started holding workshops. We put together a health questionnaire and leafleted the whole neighborhood, especially the one behind the tank farm." Some residents had been living next to the facility for three generations, so the questionnaire revealed long-term health problems. "These people were really sick. We had leukemia and other cancers. We had people with severe respiratory conditions. We had a cluster of seventy-five homes where at least one person in each home had asthma. A lot of people said that now they could relate to why they were ill and their children were ill. We told the people right away that the campaign was going to be a long campaign because we were fighting multibillion-dollar corporations. But they felt like they really wanted to take on the campaign and work. Then we had to decide what the community wanted." The residents' official requests included the following:

• A comprehensive health study of residents and former residents living near the fuel storage area.

• An East Austin health clinic to provide free medical assistance to residents and former residents in the area.

• The immediate cleanup of the soil and groundwater contamination in the area.

• The shutdown of operations of the storage facilities and the permanent removal of the tanks to a properly zoned area not adjacent to residential neighborhoods or other environmentally sensitive areas.

The requests were clear, but no headway was being made in getting anyone to do anything about them.

A state agency ran water tests that were inadequate for the potential contaminates. "They didn't do a battery of tests, only certain tests. They didn't test for heavy metals. They were testing for stuff that had already evaporated." Frustrated, one local resident hired an independent company to test his well water because when he put it on his lawn, the grass died. The test found the well contaminated with petroleum byproducts. The state agency was forced to conduct further tests. "And sure enough, when they started drilling wells, they came up dirty, dirty, and dirty." The report found contamination in 71 of the 116 water wells tested. One well had a level of benzene that was 7,100 times the federal safety limit for drinking water. The contamination was found to have spread beyond the fifty-two-acre tank farm into a local creek and five other neighborhoods in the area.

Meanwhile, community meetings were going on every week. "There was community protesting going on. There was something in the newspaper practically every day. The whole town was buzzing about it. There wasn't anyone who didn't know about it. But the city officials tried to pretend it wasn't happening because they were afraid that they were going to be sued in the whole process," says Almanza. The group went to the state air board, the attorney general's office, and the governor, but it was the county commissioners who finally acted. In May 1992, "they approved $200,000 to do a civil and criminal investigation of tank farm owners. And that's what really busted the case." The oil companies began signing agreements to relocate and cease their current operations in East Austin. The first company, Chevron, signed in August 1992, and the rest of the companies, except for Exxon, followed shortly thereafter.

Exxon continued to hold out. In October 1992, PODER and EAST initiated a letter-writing campaign addressed to the president of Exxon. Thousands of citizens from six states mailed in form letters. When the company failed to respond, the activists organized a city, state, and regional boycott of Exxon products. In February 1993, the activists turned up the heat, and began organizing a picket of the company's corporate office in Houston. "*CBS News* covered the issue, and they came down to cover the protest," says Almanza. The afternoon before the caravan of activists was to leave for Houston, Exxon agreed to relocate its facility. "I can't describe the feeling that came over us. The tears just ran," she says. But a few details still had to be tended to. "We had to call CBS. We had to call everybody. We were having to fax and call and tell all these people that there wasn't going to be a protest. It was a whole lot of work, but we were very happy."

The agreements allow the companies three years to relocate, and the county will assist in finding alternative sites. Cleaning up the site is scheduled to take fifteen years. Meanwhile, the people and the city are working together to decide what they want on the site once it's clean. Frequently, such sites become public parks.

One of their demands is still left unattended. "We're still trying to get health clinics where people in those areas will be able to go once a year for a checkup and make sure that they're all right."

In order to monitor the cleanup process during the next fifteen years, the makeshift, grassroots assemblage of mostly volunteers has had to establish itself. PODER received not-for-profit status in March 1994, which allows them to request grants from foundations. Before, they operated on small private donations and volunteers. They set up an office because "people were getting tired of calling us at seven different places," when they were working out of their homes, workplaces, and various other locations.

"You always have to be monitoring because the opposition will wait until things settle down and they think nobody is paying attention to do things. So we see it as a long process. It's nowhere near over." Already PODER has raised objections over a cleanup proposal to dump the treated water into a local creek. "Of course, when we got a whiff of it, we challenged it, and they decided just to drop that for now."

Meanwhile, PODER is involved in other issues of environmental racism. "This issue brought environmental racism to light here in the city of Austin," says Almanza. "Now it's become a very well-known term here, and corporations and different groups know they have to be very careful when they decide to put in some kind of facility. We're not just going to be the dumping ground for everything that nobody else wants."

Studies over the past fifteen years have found environmental racism occurring throughout the nation. Minorities are exposed to the most pollution and benefit the least from cleanup programs. A recent EPA report found evidence that racial and ethnic minorities suffer disproportionate exposure to dust, soot, carbon monoxide, ozone, sulfur, sulfur dioxide, and lead, as well as emissions from hazardous-waste dumps. An earlier study, by the

United Church of Christ's Commission for Racial Justice, in 1987, concluded that race, even more than poverty, was a common characteristic of communities exposed to toxic wastes. Not only that, but the larger the scale of the potential environmental hazard, the more likely it was that the operation was located in a neighborhood with a minority population. The proportion of minorities in the neighborhoods where the largest operations were located was three times greater than communities with no such operations. The first environmental racism study came out in 1979 and was conducted by a sociologist at the University of California at Riverside. He showed that since the 1920s, all the city-owned landfills and six of the eight garbage incinerators in Houston have been located in black neighborhoods, even though the majority of the residents of that city are white.

As a result of these and other studies, several lawsuits have been filed claiming a violation of the civil rights of the people in various communities. But proving intentional discrimination may be difficult or impossible to do.

Susana Almanza came to understand racism at a young age. Her family is Mayan Indian, and she grew up in the 1950s in a poor East Austin neighborhood of unpaved streets. Many families piped their own water into the homes and used outhouses. "From childhood I learned first-hand about all the injustices that exist." Austin was a segregated town in those days, "and it's still very much segregated," she says. At six or seven years old, she remembers going to the movies and having to sit in a separate section, because of the color of her skin. Even though she was a small child, "that told me that things were wrong."

By the time she was thirteen years old, she was marching in a furniture workers' strike. Later she worked on a lettuce boycott to support farm workers. She helped organize her fellow high school students to protest discrimination in school activities such as Student Council, band, and cheerleading. "Our class changed the system." The student action required the school to have a Mexican-American and an African-American on the cheerleading squad. With these campaigns under her belt, she became dedicated to making her community a better place to live. She went to work for neighborhood centers that emphasized political empowerment and were hubs for the exchange of information throughout the city and the country. She also became a member of the Brown Berets, a Latino group that worked on community organizing, established and ran free breakfast programs, set up a children's summer program, and worked to get kids to stop sniffing paint—a cheap, easily available, and dangerous way to get high in those days.

All the while, she has been raising four children. "A lot of youths are lost and have no direction and don't know where they are going. But we're from the Mayans, so this is our indigenous continent, and we still practice our indigenous ways. We still have our ceremonies. We still have our Aztec dances. We still pray. We're still very connected."

Almanza is connected with her family, her community, and the world. She exemplifies the environmental slogan "Think globally, act locally." At PODER, she says, "we will continue to impact the national industrial policy by studying trends and what corporations are doing. Even though we see ourselves as a grassroots organization, we will make our networks global because these industries are all over the world and we have to be connected to people all over the world."

FOR MORE INFORMATION:

Susana Almanza, PODER, 55 N. IH 35, Room 205B, Austin, TX 78702.

Citizens' Clearinghouse for Hazardous Waste
Telephone: 703-237-2249.

WHAT YOU CAN DO:

Pollutants such as petroleum byproducts, metals, plastics, and some chlorinated hydrocarbons can poison most forms of life. To protect yourself and your community, get involved. Find out what the local zoning permits and prohibits, before an industry wants to move in. Decide if you agree that's best for your community. Any area zoned for heavy industry is likely to be an area of heavy pollution. It should be located away from residences and schools. Look at the emissions the industry creates and how far they drift. See to it that there is at least a buffer zone around the facility.

But safe places to put unsafe industries are growing scarcer all the time. It makes sense to help create less demand for polluting products. When you lessen your dependence on toxic chemicals—from cleaning products around the house to fossil fuel transportation—you reduce the need for toxic industry.

SALMON HABITAT

Billy Frank, Olympia, Washington

*T*HE INDIAN TRIBES of western Washington have a legal right to half the fish in the waterways, according to an ancient treaty upheld by modern courts. The treaty, signed in the 1850s between the United States and the tribes of western Washington, granted the Indians salmon fishing rights because the fish are essential to their way of life. But the treaty was soon overridden by laws of the state of Washington. In 1974, the Indians won a battle in court to have the treaty upheld. The court named the tribes comanagers of the resource, a decision that was confirmed by the Supreme Court in 1979.

In 1980, though, a group of big businesses continued to fight the tribes' right to the salmon. Two bills were introduced in Congress to abrogate the treaty. The businesses also initiated yet another round of courtroom debates. "They had their bankroll, and they had a lot of things going on," says Billy Frank, who was representing twenty tribes of mostly unemployed, broke Indians.

He found out one of the main contributors to the opposition was the Sea First Bank. "We said, 'We're going to boycott,'" he remembers. His group didn't have any money to boycott with. But they called on a tribe in eastern Washington that had sixteen million dollars in the Sea First Bank, and the tribe agreed to pull its money out.

Frank talked with tribes in Alaska, and they pulled eighty million dollars out of the bank. "Now we didn't know that the Sea First Bank at that time was going under from some oil investments in

Oklahoma and Texas. But the next day I got a call from the president of the Sea First Bank," says Frank with a laugh. "And he said, 'Before I jump out of the seventeenth floor of the Sea First Bank in Seattle, I've got to sit down and talk with you.'" Frank agreed; he had a long list of things to talk about, and he remembers it came down to his saying, "You have to start moving forward with recognition of the tribes and the treaty in the Northwest, and lay out an agenda we can follow. Recognize this treaty is here to stay, and get away from trying to abrogate it. Be an advocate for us." The bank agreed. "That boycott got their attention. It hit them right in the pocketbook and got their attention. But a lot of things were happening at the same time. This was just one of the things that we took on to get the attention of the people who run the state—the business community, the big money guys."

At the same time, forest management and the destruction of water resources became more hotly contested in Washington and other states of the Pacific Northwest. The Indians joined environmentalists against the timber industry. State regulators stood in between, facing legal challenges and pressure from all sides. Round after round of legal and political debate touched off new debate. Almost inevitably, the timber industry called proposed changes a disaster, while the advocates of fish and wildlife called them inadequate. Soon, the only thing that was certain was more uncertainty over the management of forest lands in Washington State. By then, the timber industry was also faced with extremely poor market conditions. Loggers continued to lose jobs due to automation and exportation. And Billy Frank had been arrested ninety times for protest actions.

Gradually, all of the concerned parties reached the same fork in the river: they were exhausted from the battles and legal fees. Not the least exhausted was Frank—and the tribes he represented.

However, Frank was the one who came up with a new approach. He suggested that everyone send away the lawyers and sit down to a discussion. It was the only thing they hadn't tried up to that point, and all parties agreed, in July 1986, to find out how they might manage to "agree to agree." Forty representatives from the tribes, the timber industry, environmental organizations, and state government attended a three-day retreat. Tribal suggestions included less-damaging road cuts, preservation of trees around streams, and other improved logging techniques.

An intensive series of one hundred meetings followed, resulting in a draft agreement in December 1986 and a final agreement in February 1987, that was publicly reviewed, adopted, and unanimously implemented through the state legislature. The loggers accepted restrictions on practices that were environmentally damaging; they also allowed tribal and environmental representatives to participate in harvest plans and to suggest ways to manage the forest and river. The resulting Timber Fish Wildlife Agreement is unique in the United States and perhaps the world. It is not an institution, but a living process built on trust, commitment, and above all, cooperation.

"Rather than fighting," says Frank, "we're negotiating. Rather than suing each other, we're putting together teams and combining resources to properly manage the natural resources we all depend on." For his part, Billy Frank was awarded the Albert Schweitzer Prize for Humanitarianism in 1992, the Martin Luther King, Jr., Distinguished Service Award for Humanitarian Achievement in 1990, and several other national prizes.

Today the tribes, originally skeptical of cooperation with a state that had lied to and betrayed them so often in the past, are involved in every facet of Puget Sound water quality management efforts. The tribes bring their thousands of years of experience in meeting the harvest needs of their people while protecting the fish and their environment to a joint effort that will benefit all fishermen.

The Indians have the right to half the fish in the river, but half of nothing is nothing. "There used to be so many salmon out there, no one had a use for them all. Now there are few enough we count almost every one," says Billy Frank. "The courts won't fix that."

Frank has been chairman of the Northwest Indian Fisheries Commission (NIFC) since 1977, and before that was his tribe's fishery manager, an elected position he held for thirty years.

Billy Frank's father was the last full-blooded member of the Nisqually, a tribe that lived on the Nisqually River, which flows from the base of Mount Rainier to Puget Sound. Through thousands of years, the river spawned a cultural and spiritual strength in the people. Their customs and ceremonies reflected harmony with nature, kinship with its elements, and gratitude for the gifts of Mother Earth. The salmon in particular were sacred to the Indian way of life. Each year at the beginning of the salmon run, a ceremony was held to honor the first catch of the season.

This age-old reliance of the tribes on fish and wildlife, as well as on vegetation for food and medicine, remains today. Many of the Indian tribes of western Washington depend completely on the salmon for food and livelihood. The reduced salmon catch deeply hurts Indian communities where unemployment rates are among the highest in America—often approaching ninety percent. These tribes who depend on salmon for food and livelihood are standing up for the environment as well as for their rights.

The salmon in the rivers of the Pacific Northwest once thrived in legendary numbers. As recently as 1972, a *National Geographic* piece described a sockeye salmon migration as resembling a traffic jam roiling the river. People said there were so many fish in the water that it looked as though you could walk across their backs to reach the other side. Now, the salmon habitat has been damaged, migrations are hampered, and the salmon have been overfished.

The unique life cycle of the salmon requires them to swim thousands of water miles with environmental obstacles all along the way. At the beginning of the cycle, salmon eggs are laid in shallow, freshwater streams or riffles off the streams. The young salmon live in fresh water for about eighteen months, then swim downstream to the sea, where they feed for anywhere from one to five years and may grow to five feet long or more, and weigh as much as one hundred pounds.

At the proper time, driven by powerful instincts, they leave the ocean to swim back to their birthplaces to spawn. They locate the same river and ascend it, frequently up high mountain slopes, to the same stream, sometimes even the same riffle, where their life began. They seem to make their way by smell, their delicate nostrils detecting the odor of their infancy, which may be diluted one part to a billion.

A female lays thousands of eggs, deposited among aquatic vegetation or buried in the sandy bottom. The Pacific salmon species die after they spawn. (Atlantic salmon, however, live after breeding, return to the sea, and may come back upstream to spawn a second time.) Most of the thousands of eggs and newly hatched larvae are lost to natural causes, such as being eaten by other creatures. On average in nature, only one of the young survives to replace each parent. Because the parent salmon spawns only once and produces only one offspring, and because their habitat covers such a wide range, salmon populations can be quickly reduced—and have been—by damage to waterways from logging, dams, irrigation, and pollution.

Logging caused the greatest destruction to waterways in Washington and the Pacific Northwest in the early 1970s. Loggers in bulldozers smashed right through streams and creeks. In the practice called clearcutting, all the greenery, tall and small, was stripped from mountainsides. Even the streambanks were cleared of brush and vegetation. The bare mountainsides, without tree roots to hold the soil, slid into the rivers and streams in the valleys, smothering salmon spawning grounds and juvenile salmon. Even the waterways that escaped landslides rarely survived, because without their protective tree cover they soon began to die from sunburn. Meanwhile, the stripped streambanks eroded into the streams, causing them to become shallower and warmer, and unfit for many fish, their eggs, and other essential organisms. Clearcutting still goes on today.

Dams create another dramatic transformation of waterways. Intended to harness and control rivers, they choke and strangle them. Not only does their physical presence prevent or interfere with salmon migration, but dams also change the temperature of the water and the life of delicate salmon eggs, fingerlings (baby fish), and the organisms they feed on.

The list of ways we have invented to mismanage waterways goes on—as does the damage to salmon populations. Altering or straightening a river or stream to make way for development can eliminate spawning grounds. Diverting water for cities' needs and agricultural irrigation lowers the water levels of the whole waterway, as well as those of its bays and sounds. Lower water levels may cause salmon birthplaces to dry up. Pollution at the headwaters can damage the delicate eggs and other fragile organisms.

"We can't undo some of the things we've done," says Frank. "But we can bring the fish back, if we get the people behind it. And if we don't, the salmon are gone. We won't be far behind. Our survival as Indians depends on our linkage with our past. There is much we can learn through the wisdom of the ages. Indian and non-Indian alike must listen to the lessons passed along through the generations, lessons that kept this land and its resources healthy and pure from time immemorial."

FOR MORE INFORMATION:

The Northwest Indian Fisheries Commission, 6730 Martin Way E., Olympia, WA 98506.

Telephone: 206-438-1180.

WHAT YOU CAN DO:

Billy Frank reminds us that interacting with nature is a sacred gift. Even in our busy lives, we can take the time out every day to observe a flower or look up at a patch of night sky. Look into a pond or stream or bay and remember that water is a finite resource, and competitive demands for it easily exceed its availability. Practice water conservation. Fix leaky faucets—they can waste gallons of pure, fresh drinking water. If you can, install low-flow toilets that use 1.6 gallons of water per flush and showerheads that flow at 2 gallons per minute.

RUNNING WITH WOLVES

Renée Askins, Moose, Wyoming

OLVES ALL BUT VANISHED from the landscape, systematically obliterated under the banner of "predator control." Millions of wolves were not just killed but tortured to death. They were lassoed and torn apart by their limbs. Their jaws were wired shut, and they were left to starve. They were doused with gasoline and set on fire. For many people, just to hear the eerie howl of wolves brought up horrific sensations of the cruelty of nature. It is true that a wolf, with its many sharp teeth, some of them very long, can tear apart a carcass in seconds, but there is no record in North America of any wild wolf killing or seriously injuring a person—nor have they destroyed much livestock, either. Most of the wolves left in the United States live in Alaska, where they are hunted with semiautomatic weapons from helicopters.

Wolves are socially sophisticated and function in families in which all members contribute. Living in packs, they learn from, and about, and with others. They communicate well, and when the pack senses prey, the wolves rush to touch each others' noses. Disputes within the pack members are rarely settled with fights.

The Endangered Species Act has mandated the return of wolves to Yellowstone National Park, where they had become extinct. But there are still no wolves in Yellowstone because formidable opposition prevented their reintroduction. The strongest opposition has come from organized groups of ranchers and hunters, as well as the politicians who depend on their votes. The ranchers worry about wolves killing their livestock, and the hunters worriy that wolves will cut into the large game herds of elk, moose, and deer. There is some validity to both points. Federal biologists estimate that a healthy population of Yellowstone wolves will kill, each year, an average of nineteen cattle, sixty-eight sheep, and twelve hundred deer, elk, and moose.

To help relieve the worries of the ranchers, a conservation group called Defenders of Wildlife has set up a $100,000 fund to pay for livestock lost to Yellowstone wolves. As for the wild herds, Renée Askins, executive director of The Wolf Fund, says that in Yellowstone they have increased by as much as eighty percent in the past twenty years and need natural culling. Rather than hurting the ecosystem of the park, wolves could help restore its balance.

Wolves could also help the economy of the park, too. The U.S. Fish and Wildlife Service projected the economic impact of returning wolves to Yellowstone and central Idaho and found there would be a net gain of $10.4 million. Surveys show that more than $20 million would be generated annually by additional visitation to Yellowstone, if people knew wolves lived in the park. The cost of the reintroduction is about $6.7 million. Losses to the livestock and hunting industries are estimated to cost a maximum of $2.9 million.

As much as the park needs the wolves, the wolves also need Yellowstone, one of the few places in the forty-eight contiguous states where wolves might find enough habitat and prey. The park stretches across 2.2 million acres of virtually unspoiled wilderness—primarily remote, mountainous terrain—from northwestern Wyoming into southern Montana and Idaho. "Wolves belong in Yellowstone," says Askins. "You cannot argue that. It is the world's oldest national park, and it has every plant and animal species that existed when Europeans hit our shores—except wolves. Yellowstone is the largest temperate ecosystem in the world, a World Biosphere Reserve. Its management techniques inform and guide parks throughout the world. It should be a model." Yellowstone without the wolf, she says, "is like a watch without a mainspring. Like a heart without a heartbeat."

Many people agree with the goals of The Wolf Fund. A recent study showed that a six-to-one majority of visitors to Yellowstone felt that the presence of wolves would improve the Yellowstone experience. A majority of residents of Wyoming and Idaho also favor reintroduction.

In May 1994, the U.S. Fish and Wildlife Service produced a new plan to return wolves—a plan that may be satisfactory to both sides. It calls for wolves to be reintroduced to Yellowstone and federal land in central Idaho and northwestern Montana, but if they wander onto private lands and threaten livestock, ranchers can shoot to kill. "We didn't get everything we wanted," says Askins, "but the cowboys didn't either. This plan will protect wolves and people."

Producing the plan, she says, "really was a monumental achievement in bringing together disparate factions. Although the process has been very long and certainly defined by a lot of manipulation and sabotage, I think in the end the process has really been a very fruitful one in bringing these factions together and actually forging a solution that has been derived from, rather than imposed upon, the region."

Askins feels that, "On the deepest level, the issue of Yellowstone wolf recovery is not about wolves. It is about control of the West. The return of wolves, after the livestock industry exterminated them, symbolizes a shift in that control."

Whenever a federal action is proposed that may have an impact on the environment, an Environmental Impact Statement (EIS) is required. To write the EIS, a team of experts first determines what concerns should be addressed. They then hold open meetings in communities throughout the affected region. In the case of the wolf recovery plan, they also held open meetings in cities around the nation as well. Comments gathered from citizens guide the EIS team in preparing the first draft of the EIS, which is released for further comment. Public hearings are often held during this period. Then the EIS team writes a second, final draft of the EIS. In the case of the EIS for the Yellowstone wolf recovery, by the time the final drafting stage was reached, the team had received 160,000 comments—more than anyone can remember for any other EIS. The final EIS offers several alternatives but recommends one as the best plan and is presented to the Interior Secretary who has the power to select an alternative other than the recommended one or to completely ignore the recommendations.

At this writing, the final EIS recommending reintroduction of wolves to Yellowstone as an experimental population has been given to Interior Secretary Bruce Babbitt. On August 16, 1994, the sixty-day public comment period opened. The opposition still has opportunities to demand more reports, more hearings, more comments, more analysis, and to file suit against the plan or to pressure the Secretary. A number of lawsuits have been filed to stop, or just to stall, the procedure, which is racing against the clock to reintroduce wolves in the winter of 1994.

Winter is the best time to relocate the wolves for several reasons. Deep snowfall will help keep them from traveling too far and will encourage them to settle into their new home. In winter, herds of deer and elk congregate, making it easier for wolves to feed. Another reason is that mid-January through February is wolf breeding season. "The whole idea," Askins says, "is that if wolves have something else on their mind besides traveling, it may help them stay put."

Askins feels very hopeful that the reintroduction will occur in

1994, but she predicts that "to the absolute eleventh hour there will be every effort to stop it." An example of the determination of the opposition occurred a few years ago when a single wolf in Montana had to be relocated to Glacier National Park. "The helicopter with the wolf was already in the air flying up when the crew got a radio message from the governor that they couldn't release the animal." In the Yellowstone case, Askins says, political pressures will surely be brought to bear, as 1994 is an election year. However, President Clinton and Interior Secretary Bruce Babbitt both publicly favor the new plan, and Babbitt has said he thinks it can withstand lawsuits.

As if this weren't suspense enough, all of the money necessary to pay for the reintroduction has not been raised. "We've still got plenty of challenges ahead of us," Askins says, with a rueful chuckle. "If the process is delayed beyond the optimal time to release the wolves, the effort will have to shift focus to the following winter. The delay will allow the opposition time to gain strength.

"Every time we think we're close to the finish line, another bomb goes off. By no means are we uncorking the champagne, but I'm very, very hopeful and I have this fluttering of just a sense of possibility. It almost seems like a dream at this point."

If all goes well, in November of 1994 some fifteen wolves will be brought down from Canada and for two or three months will be kept in enclosures in Yellowstone to make certain the animals are healthy. After they are released, in January or February of 1995, the population will grow through natural reproduction until about one hundred wolves live in the area, by about the year 2002.

Wolves live in cooperative social groups of about eight to twelve members, called packs. They howl to communicate with one another and to mark the boundaries of their territory. They have many other methods of communication, and generally the individuals in the pack get along very well.

If an area is abundant in prey, a wolf pack's territory may be as small as forty square miles, while in an area where prey is sparse, a pack's territory may be as large as fifteen hundred square miles. Wolves usually defend their territory by keeping out other wolves; they may also attack intruding dogs or coyotes.

Wolves mate for life, and a litter of four to six pups is born each spring and is tended by the entire pack. Some wolves leave the pack as young adults to find mates and start new packs. They may move into a neighboring area, or travel hundreds of miles away. A small number, perhaps 10 percent of a wolf population, may leave the pack altogether and become "lone wolves."

The adult gray wolf, also known as the northern Rocky Mountain wolf, weighs eighty to one hundred pounds and measures about six feet from the tip of its nose to the tip of its tail; males are slightly larger than females. The average adult wolf eats about nine pounds of meat per day, or about two adult deer per month, which they hunt mainly at night. Wolves have keen eyesight and can run as fast as forty miles an hour during a hunt. They have forty-two teeth, four of them very long, to pierce tough hides. They generally live about ten years in the wild. Their enemies are mostly people, although grizzly bears who discover a den may kill the pups.

Wolves spend most of their time in valley bottoms, where prey is concentrated, but Askins says they like high places, too. "They move at the edges of things," she says. "At the edge of forest and meadow, between a walk and a run, between light and dark, stillness and movement." To her, even their tracks in the snow are like a beautiful, dotted ribbon.

Askins grew up loving the outdoors, a horse-crazy girl who rode ponies in the meadows and forests near a small town in Michigan. In 1977, she entered Kalamazoo College in Michigan to study humanities, and she wrote a paper for a theology class on the role of the wolf and its Satanic association in religious tradition. Fascinated with the wolf, she switched her studies to biology, and by her junior year she was camped out in a wolf-observation hut in Indiana, where she watched three successive litters of pups raised in captivity. One day she was handed an orphan pup, forty-eight hours old. Askins cared for the pup, feeding her, cleaning her, and sleeping with her. "I kept a journal about her but the whole time, I felt like she was keeping her own journal about me. She had such dignity, and there was a level of sophistication and communication." When Askins went back to college, she sent the four-month-old pup to a behavioral lab in Minnesota. Two years later Askins, learned that the wolf had been destroyed as a precaution during a rabies scare at the lab. Because the wolf pup had given so much to Askins, she says, "it sounds silly, but I felt I had to give something back." After graduation, Askins moved to Jackson, Wyoming, to assist a biologist working on the reintroduction of wolves to Yellowstone.

A few years later, she started the nonprofit Wolf Fund, and went back to school, to get her master's degree at the Yale School of Forestry and Environmental Studies. After graduation, in 1988, she returned to Jackson, and she brought a boyfriend, folksinger Tom Rush, with her. They are still partners today, and between performances, he serves on the board of directors of The Wolf Fund, which has become a major conservation organization that has reached over eighty million people in a single year. For her advisory board, Askins has assembled some of the best-known environmentalists and environmental celebrities: Robert Redford,

Harrison Ford, Ted Turner, Peter Mathiessen, George Schaller, and Yvon Chouinard, to name a few.

Having built such an effective, well-known organization, most people would be interested in seeing it continue, but Askins insists that the day wolves are reestablished in Yellowstone is the day The Wolf Fund goes out of business. She has kept The Wolf Fund, as she intended for it to be from the start, "a single group devoted to a single project. A conservation SWAT team."

"This is a process of social change and it's a very painful process in the West. A lot of the process has very little to do with wolves. It has to do with how the West perceives itself. So what the wolves have represented is really a symbol of that change. The resistance to the wolves has been in many ways totally disproportionate to what the wolf itself represents in terms of its return, but it's always the symbol that we work with. And I really feel strongly that is precisely what the wolf embodies for so many of both the supporters and opponents. I think the result of that long, grinding process of shifting the bedrock of a value system results, in the end, in a foundation that supports far more than just the presence of wolves."

FOR MORE INFORMATION:
The Wolf Fund
Box 471, Moose, Wyoming 83012
Telephone: 307-733-0740

WHAT YOU CAN DO:
Until the very minute the wolves step back into Yellowstone, The Wolf Fund needs public support and contributions (which are tax-deductible). The organization survives solely on donations from individuals, foundations, and corporations. At the last stage of the

effort to return wolves to Yellowstone, contributions are particularly important to help pay for the physical costs of buying, transporting, and maintaining the wolves before release. You can send contributions to the address above and help pay for a plane ticket for a wolf or a good meal in its new home. If you like, telephone first, to find out the status of the project

SAVE THE TIGERS

Ullas Karanth, Nagarahole, India

*S*o much massive size and power emanates from the tiger when you come upon it in the wild, that its five-hundred-pound body not only fills your sight, but seems somehow to fill the whole forest around it as well. If the tiger roars, the sound splits the air.

Ullas Karanth, who has seen hundreds of wild tigers, says, "It's always like the first time. You see these animals in the wild, and it's very exciting. It's something that truly satisfies you in every way. It's a sense of awe at the sheer beauty and grace of the animal. The sense of power. The sense of perfectness. If you set out to design the most beautiful animal in the world, the most graceful, the most powerful looking, you'd end up with a design like this. A tiger."

Dr. Karanth, one of India's leading tiger experts, oversees the Wildlife Conservation Society's efforts in India to help save the tiger, one of Earth's fastest disappearing species. Karanth thinks there may be only three thousand to five thousand tigers left on Earth—ninety-five percent fewer than at the turn of the twentieth century. Originally, there were eight subspecies of tiger. The Java, Caspian, and Bali tigers are now extinct. The remaining five subspecies—South Chinese, Siberian, Sumatran, Indo-Chinese, and Indian—face grim circumstances.

Most of the remaining tigers today are the Indian, or Royal Bengal, tigers living in sanctuaries in India. Karanth works in the Nagarahole National Park, in southern India, where some fifty wild tigers live in 646 square kilometers protected by 250 guards. The guards protect the tigers from poachers for whom every part of the tiger is valuable—except the roar of the living beast. A strikingly patterned fur coat sells for as much as $15,000. An intact forelimb can bring in hundreds of dollars per pound. Tiger penis soup sells for $320 a bowl in Taiwan. The bones, claws, and eyes—even the whiskers—command high prices for use in Oriental potions, elixirs, tonics, and balms based on recipes thousands of years old. The newly affluent people in Asia's rapidly growing economies can afford the exorbitant products. To supply the trade, the world's last tigers are being illegally trapped, poisoned, and shot, then smuggled across international boundaries.

International efforts to save tigers have been hampered, in part, by a lack of solid data about the animals, Karanth believes. Though tigers, with their big, round, ruffled faces that almost smile, have lived on earth for hundreds of thousands of years, they have eluded study in the wild. It is their custom to avoid people. They slink through the jungle by day, camouflaged by their tawny coats and black stripes, which mimic the shifting patterns of sun and shade. They hunt under cover of darkness. "By far, most tiger conservation effort has been divorced from good science in the past. So we have not been able to monitor our conservation efforts effectively. If we do that right now, and come up with the political will, I think there is still time to save the tigers. I think we have about five to ten years to get our act together, act rationally, and we can still save tigers in some parts of the world."

Karanth received his Ph.D. in biology at the University of Florida and is said to have been hand-picked by noted wildlife biologist Dr. George Schaller, Director of Science for the Wildlife Conservation Society, to oversee the organization's efforts in India to help save the critically endangered tiger. The plan was to begin an all-India survey of tigers to better determine their numbers and needs.

"When you begin to study a tiger, you don't necessarily go and start watching tigers," says Karanth. "What is key to understanding tigers or lions or any big cat, or any big predator for that matter, is to understand the prey. Tigers don't eat big trees or grass. They live by killing animals. Tigers don't live in isolation from their habitat. They live in a complex community of other predators, including leopards and wild dogs. So when I began my work in 1986, my first goal was to look into the prey base of the tigers."

Tigers hunt at night when other animals, such as deer or monkeys, come out to eat grass, leaves, and shrubs. Good swimmers, unlike most members of the cat family, tigers will cross rivers, if necessary, to search for prey, which they don't chase, but stalk and spring upon. Tigers may make twenty tries before they kill. Their claws, massive shoulders, and forelimbs can clutch and ground an animal much larger than they are. Tigers will take a one-ton, wild oxlike animal called the gaur, the largest hoofed animal in Asia, but they tend to avoid elephants, large buffalo, and bear, the few land mammals capable of overpowering a tiger. Once the prey has been brought down, the tiger's daggerlike fangs quickly slash its throat or rip its spinal column. A large tiger stretches ten feet from nose to tail, weighs five hundred pounds or more, and devours some sixty pounds of meat a night, the equivalent of about fifty deer or boar every year. A female bringing up young will need more—about seventy to seventy-five kills per year.

"So when you're talking of a large, high-density tiger population, you are talking in terms of preserving a fairly high-density prey base," says Karanth. He began to estimate how much prey was available in Nagarahole National Park, "a paradise in a tremendous sea of humanity. Ninety-six percent of this country is earmarked for people. Only four percent for parks and animals. On that four percent there must be no compromise." The park's dense deciduous forests are probably one of the finest habitats on Earth for tigers, which typically prefer just this sort of damp and thickly overgrown jungle. Many other mammals—tiger food—do, too. To find out which other mammals and how many, Karanth and his team "walked about five hundred kilometers every year on the trails, observing animals on either side, taking precise measurements and counts." Walking in tiger territory isn't all that dangerous, "generally," he says. "They are very timid animals and tend to avoid people on foot except in rare cases. What we're really more worried about is elephants. That's really what's dangerous when you are walking."

Using sophisticated computer models generated in North American universities, Karanth's team came up with the first good estimates of prey populations in these habitats. "And what we found is that there are roughly eighty to ninety large mammal prey species available for the tiger." Nagarahole may well provide the highest density of hoofed prey species for tigers in all of Asia.

The tiger faces tough competition for this prey from the neighboring villagers. "They are going into the forests around their homes every day and killing the prey species on a staggering scale," says Karanth. "They just start shooting. They put out food bags with dynamite in them. The pigs come and bite on them and get killed and the villagers take the pig meat. They use snares. Hundreds of animals get snared every year. The snares are set for prey animals, but they also end up killing predators sometimes. A good friend of mine actually got caught in one of these snares.

"So you find these things out when you do field work . . . what a tremendous impact this kind of local hunting is having on big predators. This is not spectacular. This is not something that attracts attention, but it's happening on such a massive scale. And the antidote is very simple. The only way you can stop this is by effective patrolling on the ground. Protecting the prey automatically ensures protection of tigers. This kind of effective enforcement on the ground is often ignored in conservation recommendations. The big conservation organizations want to believe that if you have wonderful schemes that solve the long-term problems like shortage of land, shortage of biomass, educating the public . . . if you do all of these positive things, it will just turn very benign and wildlife will be protected. Unfortunately, our experience on the ground shows that while these kinds of development schemes are absolutely necessary for the long-term survival of the tiger, we have to remember, as Lord Case, the famous economist said, 'In the long run, we'll all be dead.' What's happening in the short run is this kind of drastic poaching of the prey and of tigers. This can be stopped only with the ruthless use of force. No amount of education or development work is going to stop a small percentage of the local population from doing these criminal things."

Where this force will come from is hard to say. Sixty percent of the world's tigers live in India, which has preserved an admirable number of parks for them. But setting aside preserves does not prevent poaching. The tiger populations dropped by thirty-five percent between 1989 and 1994, destroying decades of conservation efforts. Even when the criminals are caught red-handed, the Indian justice system is neither swift nor sure in meting out punishment that might deter further offenses. Cases can drag on for decades, and arrest often just delays poachers. "Western countries can give aid, political support, and money, but unless the nations having tigers wake up, the future for the tiger is not too bright," says Karanth. The park patrols "have to be given sufficient arms and the legal power to shoot if necessary. The heads of government, the higher-ups in bureaucracy and politics, really need to understand that this kind of ruthlessness is necessary."

After gathering data and coming to some conclusions about the amount of prey available to tigers in Nagarahole, Karanth says, "we started focusing on the tiger in a little more depth. But in order to do that, you need to adopt some technologies—like putting a little transmitter on its neck. Which leads to a question: How do you catch a tiger? So we merged science with traditional ancient wisdom, using a hunting method that was developed in Nepal and in northeastern India to catch tigers and kill them. It's a method that was used by royalty to slaughter tigers, bringing them to the brink of extinction. We thought it would be a good way to save them."

The method requires many people, three elephants, and bolts of white cloth about thirty-five feet long and three feet tall. A large animal is tied up as bait on a trail known to have tigers traveling it regularly. "The tiger comes by and kills the prey and drags it into dense cover." People quickly pull the white cloth into a fence around the tiger. "The person doing the darting, in this case, me, gets up a tall tree," near which a gap is left in the cloth fence.

"The elephants go on the other side and slowly disturb the tiger, push it away from the bait, which it has probably eaten most of. The tiger will try to sneak away, but it sees this strange psychological barrier, this white cloth, which it doesn't want to go near. What it ends up doing is avoiding the white cloth and coming out through the narrow opening. That is where, at a fairly close range of about fifteen meters, the animal is darted using a preloaded syringe with a precise quantity of drug. People ask me if you have to be a great marksman. You don't have to be a great marksman because you are at a close range. But you really have to hold your breath and keep cool." Once the animal is sedated, the team has about forty-five minutes to work. "That's when we go in, take body measurements, take blood samples, check it for parasites, and then put a transmitter around its neck. As the drug wears off and the animal wakes up, we don't sit that close any more." The tiger walks off into the forest with the transmitter relaying its whereabouts.

"And lo and behold, you are able to look for a needle in a haystack. But the transmitter has a very limited frequency. You still don't get your data at your desk. You still have to go out in the forest, walk miles and miles—either walk or go on elephant back or ride in cars—to try to locate these animals, hone in on them. And since they are long-lived animals," averaging a twenty-year lifespan, "data builds up daily over long periods." He smiles. "There is a notion in the public that studying tigers is all excitement. Often it is very, very dull, routine, boring work. You generate information like how much space tigers require." Karanth found that the size of tiger habitat depends on several factors such as the amount of prey and the number of other tigers in the range.

To mark their territory, tigers spray urine; the scent tells an intruder to keep off. It sends other messages, too. "Basically, the scent communication is almost like the use of telephones or calling cards. This is how tigers communicate." The scent tells a young tiger where to find its mother, and an adult male where to find a female in heat. "So there's an intricate communication system that has a major import on how many tigers can be packed into an area.

"We also found that tigers in well-protected populations have a very high mortality rate. They get into tremendous fights. It's a very competitive society." Many of the big males die of mortal wounds in territorial fights. "It's a part of nature, and it is information like this, about life and death, that really tells us how to manage these populations."

Karanth also found that tigers are very productive. "A female tigress comes into breeding age at about four, and she starts cranking out the young every two-and-a-half years." The tigress has a gestation period of 98 to 110 days before delivering a litter of two to five kittens, of which only two typically survive. The young remain with the mother until their third year, by which time, according to Karanth, the mother is likely to be pregnant again. "So they are an extremely productive species. In a protected population, you have huge surpluses. In such situations, a mild degree of poaching probably does not depress them. But if conditions are not ideal, the picture changes very radically."

Conditions are far from ideal, and the picture for tiger survival has changed radically in the past few years. But for millennia, a population of tens of thousands of tigers roamed Asia from China south into the Indian subcontinent, and from there to Sumatra, Java, and Bali; west as far as Iran and even Turkey on the Caspian Sea; and east into Siberia. "If only America had been a little smarter way back when you bought Alaska," says Karanth. "You should have bought a bit of Siberia. Then the tiger would have been an American animal, too."

In Siberia, tigers prowl unbroken forest with plenty of prey. According to the Wildlife Conservation Society, the Siberian tiger subspecies has been reduced to approximately two to four hundred animals, a pitifully small population but still enough for a good chance to survive if poaching can be controlled. But with the dismantling of the Soviet Union and the following chaos, confusion, and corruption, bureaucratic desks in Moscow view Siberian tigers as a distant concern, while poachers view them through the scopes of semiautomatic weapons. It has been estimated that in the winter of 1993-94 between eighty and ninety-six tigers were slaughtered— one quarter to one half the remaining population. The killing continues. At this rate, in two or three years any remaining Siberian tiger population will be too small to reproduce in the wild, and the fabled Siberian tiger will join other subspecies in extinction.

Overhunting extinguished the Bali tiger in the 1940s, the Caspian or Iranian tiger in the 1970s, and the Java tiger in the 1980s. The Chinese tigers are almost surely extinct in the wild. The Indian subspecies, the Bengal tiger, with approximately three to five thousand animals, is the most numerous.

The Bengal's survival is a tribute to India's conservation efforts, especially encouraged by Indira Gandhi, Prime Minister from 1966 to 1977. The tiger was declared an endangered species by 1969, and there was an official ban on the export of wild furs. Soon, all shooting of tigers was illegal. And in 1973, the World Wildlife Fund pledged one million dollars to help the government of India launch Project Tiger. Reserves were designated and have increased in number over the years to twenty-one areas. In some cases, the local populations were called upon to sacrifice. They relocated their homes, grazing grounds, and collecting sites for wood, grass, and honey.

By the end of Project Tiger's first decade, it was considered a spectacular success. Official figures showed the number of tigers inside and outside the reserves had doubled. But the numbers, sadly, reflected no reality. "Indian foresters were very good at protecting forests and were asked to count tigers," says Karanth. "Now this is a difficult task. It's a science they were not trained for. They came up with a home-grown method," that basically involved going out and counting tiger tracks in the road. Karanth has since demonstrated the method to be wildly inaccurate. However, the numbers were submitted, and "the unfortunate part was the international conservation community swallowed this hook, line, and sinker. And they repeated these numbers in the publications." In other countries, for statistics on tiger populations, they just "sat back and assumed lotus position and came up with some numbers. And all these numbers became science." The numbers lulled environmentalists into complacency.

At the same time, environmentalists failed to pay attention to the growing market for tiger parts. In the 1980s, China, which had been engaging in massacres of its own tigers for years, had a temporary oversupply for their traditional medicine market. They prescribe ground tiger bone and its elixirs to cure rheumatism, convulsions, scabies, boils, dysentery, ulcers, typhoid, and malaria, as well as to alleviate fright, nervousness, and possession by devils. Ground tiger bone is particularly prized as an aphrodisiac for men. No known studies have supported these beliefs, and in fact, most of the products sold are fake, containing little or no tiger bone or parts. "These traditional medicines are placebos, essentially," says Karanth. In addition to China, Taiwan, and South Korea also import massive amounts of tiger bone and parts for use in remedies.

As the Chinese worked through their hoards and needed to resupply, rampant poaching broke out in India and other countries. In 1993, park officials began to realize the extent of the illegal operations. A sting operation in New Delhi uncovered 850 pounds of tiger bones in a single bust. That's about forty-two tigers. In one group of apartments along a filthy alley in Delhi, the "sting operation discovered more than a dozen families engaged in the illicit wildlife trade," according to *Time* magazine (March 28, 1994). "There the once magnificent animals are skinned, their prized parts dried and packaged, and their bones cleaned and bleached. The skins travel west, often ending up in the homes of wealthy Arabs, while the bones make their way to the east, frequently on the backs of Tibetans who ferry the contraband across mountainous, sparsely populated terrain to the Chinese border."

When "the colossal magnitude of the poaching of tigers burst on the world scene," says Karanth, investigations into the crimes immediately revealed another stark fact: that earlier population figures were inflated. "These numbers were pretty meaningless, and something needed to be done. We have a moral obligation not just to criticize, but also to establish different methods, so we developed this new technique called camera tracking, which involves setting up an automatically triggered camera. The cameras are placed on trails where the tigers move frequently, and at night the tiger comes by while we are fast asleep and peacefully snoring and takes its own picture. Now, the stripes on each tiger are unique. No two animals are alike, and this is a definite, categorical way of identifying individual tigers. Probably even an eight-year-old kid can read the markings, which are very, very different and distinct. It works even when the same animal comes in at very different angles." With this technique, Karanth has begun to develop more realistic population figures to provide solid evidence for the international effort to save tigers.

An international agreement that bans buying or selling tiger parts has been in place since 1975, imposed by the Convention on International Trade in Endangered Species (CITES). In 1993, CITES warned China and Taiwan to shut down their black markets trading in tiger parts or face trade sanctions. When they did not comply, the United States took historic steps. In July 1994, President Clinton imposed trade sanctions against Taiwan for its role in illegally marketing parts of tigers and rhinos. (Rhinos are every bit as endangered as tigers. The rhinos are slaughtered for their horn.) Such action comes under provisions of the little-known Pelly Amendment, a section of the U.S. Fishermen's Protective Act that authorizes the use of trade sanctions against countries whose practices injure endangered species. This is the first time ever that Pelly Amendment penalties have been imposed.

To support efforts to combat poaching and to protect tigers and rhinos, the White House announcement called for an immediate ban on the import of some products from Taiwan. The banned products are, appropriately, those made of animal parts: the twenty-five million dollars' worth of snakeskin shoes, leather handbags, and mussel-shell jewelry that Taiwan has been exporting to the United States. All of these products are now subject to import embargoes. Sanctions against other products are threatened if Taiwan fails to halt the trade in tiger and rhino products.

Without the trade sanctions, many believe the tigers don't stand a chance. Even with sanctions, the hope for survival of the wild species is slim. Illegal poaching will continue as long as the tigers' golden hide is valued like golden nuggets. In fact, the poachers may be enriched by the sanctions increasing the rarity of tiger parts. This is an irony sometimes created by crackdowns on crime, such as gun-control laws that enrich illegal gun dealers, for example, and not generally considered a serious counterargument to effective action.

The rate of extinction of species on Earth today may equal or exceed the rate of sixty-five million years ago, when much life on earth, including dinosaurs, was wiped out by a global cataclysm, probably a giant meteor. Today the extinction is being caused by human beings. "As we get into the age of biotechnology, many of these species are going to be very useful to us," for as yet undiscovered uses, such as food and medicines. "So it's not really sentiment or acknowledgment of the fact that the tiger as a species has a right to exist, a moral right to exist. It is also in our own self-interest that we have to make an effort to save the tiger."

But more than that, Karanth says, "We share this planet. The tiger does not belong to any nation. It's a global resource, just like carbon dioxide or water. It's basically a symbol of our efforts to preserve plant and animal life, all sorts of biodiversity on Earth. Essentially, we have to remember that all life on Earth fills in our biosphere, like the thin skin of an orange, which is the globe. So, if one part of it gets infected and spoiled in some manner, the disease will spread. Loss of the tiger does not necessarily mean just that the tiger is lost. What it indicates is a cascading series of species extinctions affecting other forms of life. In many ways, it is the most serious environmental issue. We can clean water or air, but we can't replace species."

FOR MORE INFORMATION:

Global Tiger Campaign, Wildlife Conservation Society, Bronx, NY 10460-1099.

WHAT YOU CAN DO:

In American elected offices, each letter received from a constituent is deemed to represent the equivalent of two hundred votes and is taken seriously. Postcards or phone calls count almost as highly, although form letters don't count for much at all. You can influence the international effort to save tigers by writing a brief letter or postcard expressing your concern about the fate of the species and your desire that it be saved. Here are two of the most important people to write:

President Clinton, 1600 Pennsylvania Avenue, Washington, DC 20500. Congratulate the President for his historic decision to ban the importation of wildlife products from Taiwan. Thank him for his strong interest in persuading consumer nations to finally stop the illegal trade. Encourage him to focus on South Korea and to continue monitoring progress in Taiwan and China.

Bruce Babbitt, Secretary of Interior, 1849 C Street, NW, Washington, DC 20240. Thank him for his strong interest in saving the tiger by trying to persuade consumer nations to stop the illegal trade in tigers. Express your concern for the conservation of tigers.

You can also help save other endangered species by working with groups like the Earth Island Institute (300 Broadway, Suite 28, San Francisco, CA 94133-3312) and the Oceanic Society (Fort Mason Center, Building E, San Francisco, CA 94123). They need volunteers in the field to help scientists conduct their research.

INTRODUCING GREEN MEANS: A GUIDE FOR EDUCATORS

Bill Pendergraft and Dave Smith

How to Use These Materials

Green Means documents the work of environmental heroes. Eighteen videos and the case studies in the book offer models for action. These environmental heroes are methodical and practical, and they investigate exhaustively. They are solving environmental problems in practical, economically sound ways.

As you know "green" isn't simple. It has shades of gray. Pure good and bad are rare in the complex balance between the economy and the environment, and environmentalism has its costs and tradeoffs. After you've met the environmental heroes in the videos, you'll have more questions than answers, and your thinking will be more complicated, not simpler.

When used in a school, university, community college, park, museum, science camp, environmental group, or informal educational program, these materials are most effective when you have thought through the following:

- what your program is intended to accomplish
- which of *Green Means* major environmental themes you want to explore
- the educational activities and information gathering you and your students can do.

What Is Environmental Education to Accomplish?

Each environmental educator's goals are unique. However, some nationwide trends are emerging. Despite past educational successes in recycling and other beneficial practices, overpopulation continues to stretch our limited resources, and the list of environmental challenges is mounting. It is inevitable that dwindling resources will eventually force compromises in our quality of life. In this regard, the important question is not whether we will have to accommodate ourselves to diminished resources. It is only a question of how graceful the accommodation will be. That grace will depend upon the creativity and skill we instill in young people. The emerging challenge to environmental educators is to find new ways to help young people become the creative and skilled environmental managers we need. It has been suggested that our only environmental issue is education.

More and more environmental educators are responding by leading students to investigate local issues, immersing them in some of the real and practical issues we face. They have recognized that we need to understand federal, state, and local environmental management, and we need to find our roles as individuals. It is vital to create a proving ground for investigators and an incubator for the thinking skills they need. To teach in this way, environmental educators are spending less time giving answers. Instead, they're framing questions and pointing to information sources. And students are becoming more actively involved, less like classroom spectators.

In this context, environmental education weaves together disparate topics and brings them to life. It is not a separate subject. Certainly, to judge whether the environment is ailing, students need to learn the science of air, water, land, and ecosystems. But it doesn't end there. To work locally at environmental management, we also use mathematics, communication skills, and study history, government, and economics. It's no wonder that the National Research Council's education standards contain a Science and Societal Challenges section that stresses connecting students with their social and personal world. The aim is to help students fulfill their obligations as citizens.

Finally, environmental education should give students ways to become part of the solution, and it should inspire them to act. Each *Green Means* environmental hero has devised a solution to a local or global problem, and each video suggests what you can do. The stories touch on problems, but they also focus on the creativity of individuals, communities, and businesses. They raise the hope that unsolved problems may yet be solved.

Green Means Themes

The *Green Means* videos and the case studies in the book profile businesses, communities, and individuals protecting the environment. They explore the following eight environmental themes: environmentally-friendly industry, protecting biodiversity, saving endangered species, using alternative types of agriculture, waste management, protecting urban environments, conserving our resources, and cleaning up toxic waste. The available video programs are grouped by theme as follows:

Business, Agriculture & Industry
    Colored Cotton
    Barnyard Biodiversity
    The Green Cowboy
    The Green Architecture of Greg Franta
    Prairie Prophet
    Seeds of Life
    Guru of the Old Growth

Waste Management
   Sewage Sanctuary
   Surfers Ride the Eco-Wave
   Neighborhood Cleanup
Green Cities
   Urban Jungle Gets a New Spin
   Big City Greens
Conservation
   Creek Kids
   Long Island Sound Keeper
   Habitat Forming
   Salmon Habitat
   Less is More
   Here Today

Match your type of activity to these categories, and refer to case studies in the book.

Activities and Information Gathering

The *Green Means* videos and case studies show what can be done, and they are most valuable as preparation for environmental activities. For example, if you plan to become a Stream Watch site, Creek Kids can show your students part of the picture: how a high school class cleaned up a creek. Hundreds of different tried and tested environmental activities are available in dozens of books, and each year in most areas, government, business, and environmental groups offer many workshops on conducting them. The activities are of primary importance, and the *Green Means* videos and case studies can help set the tone and start the discussion.

Environmental information sources abound, but information

about your local area is most important. Your local county and city governments are usually an untapped but rich source of relevant and timely environmental facts including monitoring data, maps, water quality, permits, and environmental impact statements. Contact your city or county planning department first, and move on to higher levels of government and local industry from there.

Environmental Education Resources

Some available environmental activities are especially appropriate for use with *Green Means*. For a comprehensive approach to exploring the human impact on your local environment, an excellent resource is "Investigating and Evaluating Environmental Issues and Actions: Skill Development Modules" from Stipes Publishing Company, 10-12 Chester Street, Champaign, IL 61280 (ISBN# is 0-87563-418-4). Other sources of training and activities are:

The Center for Environmental Education. Telephone: 310-454-4585

The National Association for Humane and Environmental Education. Telephone: 203-434-866

The National Consortium for Environmental Education and Training. Telephone: 313-998-6726

The North American Association of Environmental Educators' Publications and Membership Office. Telephone: 513-676-2514

The Western Regional Environmental Education Council (WREEC), Telephone: 713 622-7411. WREEC oversees two nationwide programs providing workshops with accompanying activities: Project Learning Tree (Kathy McGlauflin, Telephone: 202-463-2468) and Project Wild (Telephone: 303 444-2390)

The *Green Means* videos and other media to support environmental education are available for educational use from:

Environmental Media

P.O. Box 1016

Chapel Hill, NC 27514

Telephone: 1-800-368-3382

FAX (919) 942-8785

e-mail: enveduc@aol.com

Working with state and provincial education and television agencies Environmental Media designs, produces and distributes media to support environmental education.

GREEN MEANS
Individual Program Descriptions

Episodes are available on videotape from Environmental Media Corporation, P.O. Box 1016 Chapel Hill, NC 27514. For details, call them at 1-800-ENV-EDUC (368-3382); fax them at 919-942-8785; contact them via E-mail at enveduc@aol.com.

101. The Green Cowboy

Wyoming rancher Jack Turnell discovered that he couldn't turn his back and ride away from the overgrazed land in this cattle-rich country. Today, he's finding new ways of grazing cattle that preserve native grasses and leave intact the crucial river zones on the High Plains.

102. Colored Cotton

Natural, vibrantly-colored cotton in green, beige, and even orange! Sally Fox, a plant breeder and farmer, grows environmentally-friendly cotton in organic, living color. Not only does this cotton not require chemical processing, but the natural fibers don't fade—in fact, the colors become more intense with each washing. You'll see why clothing manufacturers now buy her richly colored fibre as fast as she can grow it.

103. The Green Architecture of Greg Franta

In the snowy town of Boulder, Colorado, some families manage to heat their homes for as little as $11 a year. Their "secret" is the energy-efficient architecture of Greg Franta, a designer in the forefront of a growing movement to create homes that are earth-wise as well as beautiful.

104. Sewage Sanctuary

Kudos to Arcata, California, for solving a large problem the green way. Instead of building a multimillion-dollar sewage treatment plant, the city transformed a dump into a series of natural ponds filled with waste-consuming plants that process the town's sewage. Odorless, clean water now pours into the ocean, marine birds flock to the site, and this innovative wetland sanctuary has become a community park for residents and visitors alike.

105. Barnyard Biodiversity

Did you know that nearly half the livestock breeds in this country are threatened with extinction? Because most American farmers raise only breeds of chickens, turkeys, cattle, and sheep that are economically "efficient," the genetic library of livestock animals is quickly shrinking. Former veterinarian Don Bixby shows us some of the "minor breeds" he's dedicated himself to preserving, and a few North Carolina farmers demonstrate how barnyard biodiversity can be good for business.

106. Prairie Prophet

Meet philosopher/farmer Wes Jackson, recipient of a "genius" grant from the MacArthur Foundation. At his Land Institute in Salina, Kansas, he and his students are studying the ancient, hardy prairie ecosystem to develop a farming alternative that can maintain valuable topsoil. For Wes, Green Means you can feed the world without destroying it in the process.

107. Seeds of Life

What's true of plants is also true of people—especially prison inmates. Give them sun and water, and they will grow. Thanks to Catherine Sneed and the Prison Horticulture Program she created at the San Francisco County Jail, prisoners sow the seeds for a productive, healthy life by cultivating a lush, organic garden. The food they grow is sold to local restaurants and donated to organizations that feed the homeless. For many of the inmates, tending the garden is a healing process that symbolizes a fresh start.

108. Long Island Soundkeeper

Terry Backer patrols the Long Island Sound in his boat, keeping an eye out for environmental warning signs in a waterway surrounded by industry and by twenty million people. For Backer, Green Means preserving the Sound as a viable commercial fishing zone and an environmentally healthy ecosystem.

109. Urban Jungle Gets New Spin

Wildlife (the animal kind) isn't easy to find in a city of eight million people. Wetlands, hardwood forests, and meadows have a hard time coexisting with New York's urban jungle. That's why Mark Matsil and a team of volunteers are restoring an oil-polluted Staten Island salt marsh, home to forty percent of the region's wading birds—and why a group of excited schoolchildren follow Matsil's lead to clean up the last remaining meadow pond on this island.

110. Neithborhood Cleanup

It's not in the guidebooks, but if you ask nicely, the Concerned Citizens of Greenpoint will take you on a toxic tour of their urban habitat, long used as a dumping ground for New York City's waste. For this small citizen-action group rallying to clean up their Brooklyn neighborhood, Green Means they're not going to take it anymore.

111. The Population Connection

To Nancy Wallace, global overpopulation is inextricably tied to every environmental ill. That's why this inspired lobbyist prowls Capitol Hill, talking to diplomats and members of Congress about the need to control the earth's human population as the first step in preserving the earth itself.

112. Always Cry Wolf

Once upon a time, the pristine wilderness of Yellowstone National Park was home to the wolf. Area ranchers, protective of their cattle, drove them out and killed them off. Meet Renée Askins, whose campaign to re-introduce the wolf to Yellowstone is on the verge of success.

113. Habitat Forming

For Steve Packard, Science Director of Illinois' Nature Conservancy, Green Means that environments once thought to be lost can be resurrected. When Packard realized that a typical midwestern prairie field in the Northbrook suburb of Chicago was once a grassland home to oak trees and unique varieties of wildflowers, hawks, bluebirds,and butterflies, he swung into action. He tells us how he painstakingly tracked down the few remaining examples of indigenous plant life and how he reintroduced them to recreate a savanna.

114. Surfers Ride the Eco-Wave

Surf's up, but so is the amount of raw sewage that pollutes the waves and closes our beaches. Meet the Southern California members of the international Surfrider Foundation, who have taken it upon themselves to clean up their native habitat. By putting pressure on Los Angeles officials, these surfers are responsible for diverting a storm drain that was spewing effluent into Santa Monica waters.

115. Creek Kids

For the students of Casa Grande High School in Petaluma, California, the polluted Adobe Creek became a rallying point for their youthful "I Care" energies. Trout and salmon that once spawned in the stream were long gone. Abandoned refrigerators and old tires stood witness to the onslaught of urbanization and public apathy.

Inspired by their teacher, Tom Furrer, these students cleaned up the creek bed, planted a canopy of trees along the banks, and even set up a fish hatchery to restock the creek. The prognosis is better than good; the salmon have returned to Adobe Creek.

116. Salmon Habitat

Billy Frank, a Nisqually Indian and a recipient of the Albert Schweitzer Prize for Humanitarianism, has been an activist since the 1960s. Over the years, he has moved from confrontation to cooperation to ensure Native American fishing rights and to restore and protect the salmon habitat of Washington state.

117. Garbage Museum

You won't find Rembrandts or Picassos in this museum—but how about half-eaten pizzas and rotting newspapers? Welcome to the Browning-Ferris Industries Education Center near San Jose, California, better known as the "Garbage Museum." It's home to an enormous wall of garbage that represents half of what Americans toss out every second. Children and adults have fun learning about garbage as they point a laser beam at an old sneaker and an egg carton to find out if they're bio-degradable.

118. Green Business

The Aveda Corporation in Minneapolis proves that business can be kind to the environment and profitable at the same time. This 320-employee manufacturer of hair- and skin-care products is a model eco-business, as founder Horst Rechelbacher energetically demonstrates. The company's packaging uses only nonchemical soy-based inks; the cafeteria has cut solid waste by some ninety percent; it's working to eliminate petroleum from their products; and, of course, it recycles.

119. Less is More

Vicki Robin and Joe Dominguez, gurus of reduced consumption, insist that the key to living harmoniously on the planet is to want—and use—less. Their proof? They live in Seattle on just $7500 a year. As Vicki says, overconsumption threatens old-growth forests more than chainsaws do and causes oil spills as surely as tankers that run aground. Vicki and Joe have dedicated their lives to getting the word out that if we lower our demand and consume less, we can literally save the earth.

120. Big City Greens

It is home to the world's largest tire-recycling plant. It has one of the most comprehensive citywide recycling programs anywhere. It even has an inspector who makes surprise checks of trash cans to make sure that recyclables aren't being tossed out with the garbage. Is it on another planet? No, it's Newark, New Jersey—pulled into the 21st century by dynamic mayor Sharpe James, who shows us how his city is changing its reputation.

121. Car-Free Living

For urban planner Peter Calthorpe, the American Dream of two cars in every garage has produced the American nightmare: clogged freeways, one-block driving, and suburban isolation. To address the car crisis, Calthorpe designs small town communities with city centers, shops, and businesses all within walking distances. Laguna West, a planned community near Sacramento, California, is an exciting example of Calthorpe's solution.

122. Guru of Old Growth

Meet Jerry Franklin, a research botanist in Seattle, Washington, whose philosophy of harvesting trees without clearcutting is making him a diplomat in the timber wars. His "New Forestry" is now being practiced by timber companies that were once at odds with environmentalists.

123. Here Today

Photographers Susan Middleton and David Liittschwager are in a race against time to preserve on film America's vanishing plant and animal species. From the whooping crane to the Presidio manzanita—only one of which remains on earth—their photographs have become emblems, not epitaphs, for America's wildlife.

124. Windpower

Wind is cheaper than gas or oil when it is harnessed to generate energy. With the recent development of a new highly efficient turbine, wind energy can now go head-to-head against fossil fuels as a cost-effective source of energy.

125. Bats

Oregon cherry farmers Tony and Betty Koch use birds and bats for an innovative but old-fashioned—and non-toxic—form of pest control.

201. Crimes Against Nature

Ashland, Oregon is home to the U.S. Fish and Wildlife Conservatory, the only wildlife forensic laboratory in the world dedicated to stopping the $2 billion trade in endangered species. Directed by Ken Goddard, the lab applies the most sophisticated techniques, including DNA fingerprinting, in the war against poachers and others who trade in endangered animal body parts.

202. Tackling Texas Toxics

Susana Almanza's battle to rid her Latino neighborhood of huge gasoline storage tanks has helped her community confront environmental racism in East Austin, Texas. Through her efforts, PODER — an acronym standing for People Organized in Defense of the Earth and her Resources (and meaning "power" in Spanish) — has convinced six major oil companies to remove the tanks that have blighted local health and landscape for the past forty years.

203. Shamans and Scientists

The traditional knowledge of rainforest healers is being brought to Western medicine by San Francisco-based Shaman Pharmaceuticals. Lisa Conte, CEO, and her team of ethnobotanists share their profits with the rainforest communities whose knowledge of medicinal plants is helping to produce effective drugs.

204. Clean Air Cabs

A taxicab company is one of a growing number of businesses that are converting their fleets to clean-burning, natural gas fuel, helping to green the nation's capital. U.S. Secretary of Energy Hazel O'Leary waxes optimistic about the future of this alternative fuel.

205. Reef Relief

The fragile reefs of Key West were undefended against damage caused by scuba divers and pleasure boats until Reef Relief was formed. Through mooring buoys, an educational program, and a campaign to have the area declared a National Marine Sanctuary, this grassroots organization has proved to be a key protector of the reef.

206. Seeds of Change

A century ago, America's plant life in America was extraordinarily diverse. Today, all but three percent of these varieties have disappeared from the landscape. Thanks to an innovative New Mexican company that collects plants and disseminates heirloom seeds, almost-lost varieties of fruits and vegetables are being planted again.

207. Seattle Spokes

For eco-commuters, Seattle is a model American city. Having invested some $17 million in bike lanes, racks, and other facilities that encourage green commuting, Seattle is proving to be user-friendly to Dr. Steve Adam, one of the ten percent of commuters who have chosen two feet or two wheels as the best way to beat the rush hour and reduce air pollution.

208. Organic Milk

Only five dairy farms remain in the Tomales Bay region of Marin County, once the prime dairy county in California. One of these is owned by the Straus family, whose innovation in producing the first organic milk west of the Mississippi is a lesson to other small farmers who want to stay on the land.

209. Nature's Collaborator

Artist Andy Goldsworthy transforms leaves, stones, snow, feathers, ice, and other natural artifacts into environmental sculpture that makes us look with new eyes at nature and our role in it. So vulnerable are his works to the processes of nature that most of them now exist only in the photographs he takes of them.

210. The Recyclers of Cairo

The Zabbaleen are a minority Christian community in Cairo whose women and children have traditionally performed the tedious work of picking through garbage heaps and recycling what is reusable. Meet Laila Kamel, who has shown the zabbaleen that their labors can be a means to another end. The schools and weaving centers she has helped establish provide an economic foundation for new business built on what others throw away.

211. Man vs. Dam

If finished, the James Bay hydroelectric dam in the province of Quebec will flood an area the size of France. Located in the last great wilderness area in eastern North America, the dam has already killed 10,000 caribou, and rotting vegetation is poisoning the waterways with methyl mercury. Leading the battle to stop U.S. utility companies from importing power generated by the dam is Matthew Coon Come, Grand Chief of the Cree Nation, who along with the Inuit, have trapped, hunted, and fished the region for centuries.

212. The Buffalo Return

The sixty million bison that once roamed the American plains were all but gone by the year 1900. So, too, was a way of life for the Native American tribes that thrived on the buffalo. Today, cattle overgraze these grasslands. Meet Fred Du Bray, a Lakota Sioux who founded the Intertribal Bison Cooperative to restore the bison to the plains, and with it, the traditional native culture that depends on the buffalo.

213. Solar Ovens

The wood used to fuel cooking fires has denuded three million acres of forest in Kenya and made respiratory infection from indoor smoke pollution that country's leading cause of illness. Meet Kenyan field specialist Monique Nditu and Daniel Kammen, Princeton physicist and assistant professor of international affairs, as they train villagers in the construction and use of an elegant new technology: the solar oven.

214. Learning from the Lacandones

Is is possible to live in the rainforest without destroying it? The Lacandon Maya of Chiapas, Mexico, have done so for centuries. Through a long tradition of what Western scientists would call intercropping, nitrogen fixation, and crop rotation, the Lacandones produce a sustainable agriculture of nearly eighty different crops—an agriculture that may be the key to saving the rainforest.

215. Save the Tigers

The stuff of poetry and myth, the tiger is rapidly disappearing. Hunted and poached to near extinction—their habitat crowded by humans and their prey decimated—fewer than 6,000 tigers are left in the wild. There is hope in Southern India, largely through the efforts of biologist Ullas Karanth, whose studies may lead the tiger to a peaceful coexistence with ever-expanding human populations.

BIBLIOGRAPHY

BOOKS

Ackerman, Diane, *The Moon by Whale Light*, Random House, New York, 1991

Allen, Thomas B., editor, *The Marvels of Animal Behavior*, National Geographic Society, Washington, D.C., 1972

Ausubel, Kenny, *Seeds of Change: The Living Treasure*, HarperCollins Publishers, New York, 1994

Bixby, Donald E.; Christman, Carolyn J.; Ehrman, Cynthia J.; Sponenberg, D. Phillip, *Taking Stock: The North American Livestock Census*, The McDonald & Woodward Publishing Company, Blacksburg, Virginia, 1994

Bonner, John Tyler, *The Scale of Nature*, Harper & Row, New York, 1969

Bormann, Herbert F., and Kellert, Stephen R., *Ecology, Economics, Ethics: The Broken Circle*, Yale University Press, New Haven, 1991

Bullard, Robert D., editor, *Unequal Protection: Environmental Justice and Communities of Color*, Sierra Club Books, San Francisco, 1994

Carson, Rachel, *Silent Spring*, Houghton Mifflin Company, New York, 1962

Caufield, Catherine, *In the Rainforest*, University of Chicago Press, Chicago, 1986

Corbett, Jim, *Jungle Lore*, Oxford University Press, London, 1953

_____, *Man-Eaters of Kumaon*, Oxford University Press, London, 1944

_____, *The Man-Eating Leopard of Rudraprayag*, Oxford University Press, London, 1948

_____, *The Temple Tiger and More Man-Eaters of Kumaon*, Oxford University Press, London, 1954

Daly, Herman E., and Cobb, John B., Jr., *For the Common Good: Redirecting the Economy Toward Community, the Environment, and a Sustainable Future*, Beacon Press, Boston, 1989

Dary, David A., *The Buffalo Book: The Full Saga of the American Animal*, Sage Books, Swallow Press/Ohio University Press, 1989

Dobson, Andrew, editor, *The Green Reader: Essays Toward a Sustainable Society*, Mercury House, San Francisco, 1991

Dwyer, August, *Into the Amazon: The Struggle for the Rain Forest*, Sierra Club Books, San Francisco, 1990

Ehrlich, Paul R., and Ehrlich, Anne H., *The Population Explosion*, Simon & Schuster, New York, 1990

Fowler, Cary, and Mooney, Patrick R., *Shattering: Food, Politics, and the Loss of Genetic Diversity*, University of Arizona Press, Tucson, 1990

Gennino, Angela, editor, *Amazonia, Voices from the Rainforest,* Rainforest Action Network, San Francisco, 1990

Gore, Senator Al, *Earth in the Balance: Ecology and the Human Spirit,* Houghton Mifflin Company, New York, 1992

Graves, T., editor, *Intellectual Property Rights for Indigenous People: A Sourcebook,* The Society for Applied Anthropology, Oklahoma City, OK

Hamilton, Lawrence S., and Snedaker, Samuel C., editors, *Handbook for Mangrove Area Management,* United Nations Environment Program and East-West Center, Honolulu, 1984

Hart, John, *Farming on the Edge,* University of California Press, Berkeley, 1991

Huysmans, J.-K., *Against Nature,* Penguin Books, New York, 1986

Jackson, Wes, *Altars of Unhewn Stone,* Farrar, Straus & Giroux, New York, 1987

_____, *Becoming Native to This Place,* University Press of Kentucky, Lexington, 1994

_____, *New Roots for Agriculture,* University of Nebraska Press, Lincoln, 1985

_____; Berry, Wendell; and Colman, Bruce, editors, *Meeting the Expectations of the Land,* Farrar, Straus & Giroux, New York, 1984

Leopold, Aldo, *A Sand County Almanac (1949),* Oxford University Press, Oxford, 1968

Myers, N., *The Primary Source: Tropical Forests and Our Future,* W. W. Norton, New York, 1984

Newsday staff, *Rush to Burn: Solving America's Garbage Crisis?,* Island Press, Washington, D.C., 1989

Padoch, Christine, and Denslow, Julie Sloan, editors, *Peoples of the Tropical Rainforest,* University of California Press, Berkeley, 1988

Pickett, S.T.A., and White, P. S., *The Ecology of Natural Disturbance and Patch Dynamics,* Academic Press, Orlando, 1985

Roe, F. G. *The North American Buffalo: A Critical Study of the Species in Its Wild State,* University of Toronto Press, Toronto, 1972

Russo, Ron, *Mountain State Mammals,* Nature Study Guild, Box 972, Berkeley, California 94701

Schumacher, E.F., *Small Is Beautiful,* Harper & Row, New York, 1973

Soulé, M., and Kohm, Kathryn, *Research Priorities for Conservation Biology,* Island Press, Washington, 1990

Soule, Judith D., and Piper, Jon K., *Farming in Nature's Image,* Island Press, Covelo, California, 1992

Stafford-Deitsch, Jeremy, *Reef: A Safari Through the Coral World,* Sierra Club Books, San Francisco, 1993

Von Frisch, Karl, *Animal Architecture,* Harcourt Brace Jovanovich, 1974

Wallace, Aubrey, *Eco-Heroes, Twelve Tales of Environmental Victory,* Mercury House, San Francisco, 1993

Ward, Geoffrey C., with Ward, Diane Raines, *Tiger-Wallahs, Encounters with the Men Who Tried to Save the Greatest of the Great Cats*, HarperCollins Publishers, New York, 1993

Wells, Sue, and Hanna, Nick, *The Greenpeace Book of Coral Reefs*, Sterling Publishing Company, New York, 1992

Williams, Joy, *Florida Keys—History and Guide*, Random House, New York, 1987

Wilson, Edward O., editor, *Biodiversity*, National Academy of Sciences, Washington, D.C., 1988

_____ , *The Diversity of Life*, Harvard University, Cambridge, 1992

PERIODICALS

Ackerman, Diane, "Insect Love," *New Yorker*, August 17, 1992

Askins, Renee, *The Wolf Fund Newsletter*, Winter/Spring 1993

Babington, Charles, "Scandal in High Places," *Wildlife Conservation*, November/December 1990

Bowermaster, Jon, "Cry Wolf," *Harper's Bazaar*, March 1994

Burton, Thomas M., "Magic Bullets—Drug Company Looks to 'Witch Doctors' to Conjure Products", *Wall Street Journal*, July 7, 1994

Caufield, Catherine, "The Ancient Forest," *New Yorker*, May 14, 1990

Dawidoff, Nicholas, "One for the Wolves," *Audubon*, July/August 1992

Eisenberg, Evan, "Back to Eden," *The Atlantic*, November 1990

Ellis, William S., "The Gift of Gardening," *National Geographic*, December 1992

Franklin, Dr. Jerry, "Toward a New Forestry," *American Forests*, November/December 1989

Franklin, Jerry F., "Lessons from Old-Growth," *Journal of Forestry*, December 1993

Hoffman, Winifred, and Hoffman, Kenneth, "Grass Dairying: New Horizons for Heritage Breeds," *The American Livestock Breeds Conservancy News*, January/February, 1994

Kammen, Dan, "Cooking in the Sunshine," *Nature*, November 29, 1992

Krajick, Kevin, "Sorcerer's Apprentices," *Newsweek*, January 18, 1993

Linden, Eugene, "Tigers on the Brink," *Time*, March 28, 1994

Luoma, Jon R., "Prophet of the Prairie," *Audubon*, November 1989

Mead, S. B., "Catalogue of Plants Growing Spontaneously in the State of Illinois, the Principal Part near August, Hancock County," *The Prairie Farmer*, 1846

Packard, Steve, "Just a Few Oddball Species: Restoration and the Rediscovery of the Tallgrass Savanna," *Restoration & Management Notes*, The Society for Ecological Restoration, Summer 1988

_____ , "Restoring Oak Ecosystems," *Restoration & Management Notes*, The Society for Ecological Restoration, Summer 1993

Ryan, John C., "Goods from the Woods," *World Watch*, July/August 1991

Simonelli, Richard, "Tatanka Returns," *Winds of Change,* Autumn 1993

Sleeper, Barbara, "Scotland Yard for Wildlife," *Animals,* September/October, 1993

Sletto, Jacqueline W., "Prairie Tribes and the Buffalo," *Native Peoples,* Winter 1993

Speart, Jessica, "War Within," *Buzzworm: The Environmental Journal,* July/August, 1993

van de Vliet, Anita, "Seeds of Success," *World Link,* 1994

Ward, Geoffrey C., "The People and the Tiger," *Audubon,* July/August 1994

Williams, Philip, "The Debate Over Large Dams: The Case Against" *Civil Engineering,* August 1991

REPORTS

Alternative Agriculture, Committee on the Role of Alternative Farming Methods in Modern Agriculture, Board on Agriculture, National Research Council, National Academy Press, Washington, D.C., 1989

Bison Cultural Traditions of the Northern Great Plains: Past, Present and Future, A Report on the Bison Cultural Presentation, June 17, 1993, at the Wind River Reservation, Wyoming, written by Lauren McKeever for the InterTribal Bison Cooperative

Forest Ecosystem Management: An Ecological, Economic, and Social Assessment Report of the Forest Ecosystem Management Assessment Team, July 1993

Lost Crops of the Incas: Little-Known Plants of the Andes with Promise for Worldwide Cultivation, report of an ad hoc panel of the Advisory Committee on Technology, National Academy of Sciences, National Academy Press, Washington, D.C., 1984

Toxic Wastes and Race in the United States: A National Report on the Racial and Socio-Economic Characteristics of Communities with Hazardous Waste Sites, Commission for Racial Justice, United Church of Christ, New York, 1987

PAPERS

Axt, Josephine R.; Corn, M.; Lee, M.; Ackerman, D., Biotechnology, Indigenous Peoples and Intellectual Property Rights, Congressional Research Service, The Library of Congress, Washington, D.C., 1993

Bolze, Dorene, Policy Analyst, Wildlife Conservation Society, The Rhino and Tiger Conservation Act of 1994, Testimony before the Subcommittee on Environment and Natural Resources of the Committee on Merchant Marine and Fisheries of the U.S. House of Representatives

Brown, Lester, The Changing World Food Prospect: The Nineties and Beyond, WorldWatch Paper, Washington, D.C., 1988

The Ecologist, Briefing Document: The Social and Environmental Effects of Large Dams, Camelford, Cornwall, U.K.

Swanson, F. J., and Franklin, J. F., "New Forestry Principles from Ecosystem Analysis of Pacific Northwest Forest," from Ecological Applications, Vol. 2, No. 3, August 1992, published by the Ecological Society of America

GOVERNMENT PUBLICATIONS

CITES, Appendices I, II, and II to the Convention on International Trade in Endangered Species of Wild Fauna and Flora, September 30, 1992, U.S. Fish and Wildlife Service, Washington, D.C.

Endangered and Threatened Wildlife and Plants, Title 50—Wildlife and Fisheries, Part 17.11 & 17.12, U. S. Government Printing Office, 1993

Endangered Species Act of 1973, U.S. Fish and Wildlife Service, Washington, D.C.

Lacey Act, U.S. Government Printing Office, 1989

Pesticides Industry Sales and Usage, 1990-1991 Marketing Estimates, U.S. Environmental Protection Agency, Washington, DC, Fall 1992

ACKNOWLEDGMENTS

Peter Stein

My least favorite moments as executive producer of *Green Means* came in the edit room when, after watching a wonderful five-minute story, I had to turn to the producer and editor and remind them that they must deliver a just-as-wonderful four-minute story. To add further insult to the hard-working staff, the brevity of our *Green Means* pieces required the elimination of traditional production credits. So it is with tremendous relief that I have the opportunity here to acknowledge the talented people all over the country, but particularly at KQED in San Francisco, who made the series possible. Since *Green Means* has stretched over two and half years and forty stories, it's a sizeable list. So, belatedly, roll the credits:

Associate Producers: Wendy Zheutlin, who researched and developed many of the stories and found eco-heroes in some of the most unlikely places; and Robin Epstein, whose good humor and indefatigable problem-solving salvaged every production crisis, along with my sanity.

Segment Producers: These are the gifted people whose work on location and in the edit room transformed ideas on paper into living stories: Ray Telles, Wendy Hanamura, Michael Lerner, Richard Saiz, David Helvarg, Ken Ellis, Elizabeth Farnsworth, Sam Hurst, Michelle Riddle, Richard Vaughan, and Jaime Kibben.

Editors: Kitty Rea, Herb Ferrette, Linda Peckham, Shirley Thompson, and Blair Gershkow, who together determined the look and feel of our stories; and John Andreini, our brilliant CMX editor, who in the course of on-line editing all forty stories had to squeeze sounds out of a ferret and re-stripe a tiger.

Videographers (they shoot the scenes): John Claibourne, Jim Raeside, Greg King, Robert Shepard, Scott Ransom, Richard Neill, Ed Matney, Tom Vlodek, Ed Fuentes, Dale Green, Stuart Keene, and Carl Hersh.

Audio and Technical Engineers: Helen Silvani, djovida, Birrell Walsh, Steve Bartz, Charles Tomaras, Sekou Shepard, Chris Hall, Rene Grebelo, Susan Plambeck, Willy Gomez, Pat Marek, Laurie Curtis, Joel Sartori, Rosalie Gancie, and all the folks at Rodel Audio and NPR.

A special thanks to our Washington, D.C. crew who made our shoots with Susan Stamberg a breeze, albeit a humid one: Vince Gancie, camera; J.P. Whiteside, audio; Bill Smith, prompter; Andrea Barlow, and Martha Bota, make-up; Ivan Katz and Tracy McDonough, grips.

The Washington locations were gorgeous: thanks to Brookside Gardens in Montgomery County, Maryland, and Airlie Center in Warrenton, Virginia, who opened their doors and outdoors to us; and to Lianne Williamson, Georgia Smith, and Betsy Kleeblatt, who scouted them out.

Designers: Jim Yager, Margaret McCall, San Francisco Production Group.

Music: Mark Adler, whose unfailing sensitivity to mood and pictures is a constant source of delight.

Unit Manager: Jolee Hoyt, who, besides keeping the project on budget, managed to find us a warm hotel in Meeteetse, Wyoming.

Our hard-working interns: Scott Carroll, Ben Dillon, Marc Largent, Wendy Leopold, Chloe Levy, Shana Levy, Mike O'Donnell, Anna Rosenberg, and Vicky Zika.

A host of people at KQED worked behind the scenes on the marketing and development of *Green Means:* Joanne Sutro, Robbie Fabian, Regina Eisenberg, Danica Michels, and Nan Hohenstein; in production and engineering, Michael Schwarz, Greg Swartz, Larry Reid, Vega Gardner, Gigi Lee, Simon Hui, Eric Dauster, Walt Bjerke, Margaret Clarke, and Peter Borg; and in administration, Margaret Berry, Nancy Fernandez, and Arlette Labat.

My deep gratitude goes to Richard and Rhoda Goldman and to the foundation that bears their name, whose outspoken support of environmental causes is an inspiration, and whose sponsorship of the series was essential, and to Duane Silverstein of the Goldman Environmental Fund for his continued commitment to *Green Means*.

And finally a big thanks to Susan Stamberg, "the radio lady," who went literally to extremes (snowstorms to heat waves) to bring her inimitable warmth and style to the series. While we may have taught her a few television tricks (chewing on ice cubes so your breath won't show in cold weather), just listening to her read is a lesson in storytelling.

ACKNOWLEDGMENTS

Aubrey Wallace

I am deeply grateful to the individuals profiled in this book—not only for their willingness to share with me, and you, their stories—but also for their hard work and commitment to helping preserve our environment. My gratitude also goes to my editor, David Gancher, for his unflagging faith in me. His advice, friendship, and emotional support over many years have been a source of great strength. Pam Byers, publisher of KQED Books and Tapes, cheerfully trusted my editorial judgment and allowed me creative freedom and expression. Peter Stein, executive producer of the *Green Means* television series, helped me fashion several of the chapters and freely shared his resources. Tim Pearson's computer magic created the striking photographs. Independent producer Wendy Zheutlin generously shared background information and material, as did independent producer Richard Saiz. Robin Epstein diligently kept me informed of all the latest stages and changes in production. Elizabeth Wright provided strong backup and enthusiasm—as well as pertinent clippings—when I needed them most. Special thanks go to my wonderful friends and family who encouraged me with warmth and respect.

ABOUT THE RICHARD & RHODA GOLDMAN FUND

Since its inception, the television series *Green Means* has been generously underwritten by grants from the Richard & Rhoda Goldman Fund, a San Francisco-based philanthropic organization supporting a variety of causes nationwide, with a particular emphasis on environmental issues.

In 1990, the Richard & Rhoda Goldman Fund established the Goldman Environmental Prize, the world's largest prize program honoring grassroots environmentalists. The prize is given annually to six environmental heroes from each of the world's inhabited continental regions. In sponsoring the prize, the Goldman Fund seeks to provide eco-heroes the recognition, visibility, and credibility their efforts deserve.

ABOUT THE AUTHOR

Aubrey Wallace's last book was *Eco-Heroes, Twelve Tales of Environmental Victory,* published by Mercury House, San Francisco, California, 1993. She was also editor of *The New Environmental Handbook,* published by Friends of the Earth, and has published widely in magazines and newspapers.

Public Broadcasting Station!

Every community across America is reached by one of the 346 member stations of the Public Broadcasting Service. These stations bring information, entertainment, and insight for the whole family.

Think about the programs you enjoy and remember most: *Mystery* . . . *Masterpiece Theatre* . . . *Nova* . . . *Nature* . . . *Sesame Street* . . . *Ghostwriter* . . . *Reading Rainbow* . . . *"I'll Fly Away"* . . . *MacNeil/Lehrer News Hour* . . . *Great Performances* . . . *National Geographic* . . . *American Playhouse* . . . and so many more.

On your local PBS station, you'll also find fascinating adult education courses, provocative documentaries, great cooking and do-it-yourself programs, and thoughtful local analysis.

Despite the generous underwriting contributions of foundations and corporations, more than half of all public television budgets come from individual member support.

For less than the cost of a night at the movies, less than a couple of months of a daily paper, less than a month of your cable TV bill, you can help make possible all the quality programming you enjoy.

Become a member of your public broadcasting station and do your part.

Public Television. You make it happen!

PBS